Methods of Molecular Analysis in the Life Sciences

Delivering fundamental insights into the most popular methods of molecular analysis, this text is an invaluable resource for students and researchers. It encompasses an extensive range of spectroscopic and spectrometric techniques used for molecular analysis in the life sciences, especially in the elucidation of the structure and function of biological molecules.

Covering the range of up-to-date methodologies from everyday mass spectrometry and centrifugation to the more probing X-ray crystallography and surface-sensitive techniques, the book is intended for undergraduates starting out in the laboratory and for more advanced postgraduates pursuing complex research goals. The comprehensive text has a strong emphasis on the background principles of each method, including equations where they are of integral importance to the individual techniques. With sections on all the major procedures for analysing biological molecules, this book will serve as a useful guide across a range of fields, from new drug discovery to forensics and environmental studies.

Andreas Hofmann is the Structural Chemistry Program Leader at Griffith University's Eskitis Institute in Brisbane, Australia, and an Honorary Senior Research Fellow in the Faculty of Veterinary Sciences at the University of Melbourne. His research focuses on the structure and function of proteins in infectious and neurodegenerative diseases.

Anne Simon is an Associate Professor at the University of Lyon in France. Her research focuses on material biofunctionalisation, biomaterials, cellular adhesion, supported or free-standing lipid membranes, the study of biological membrane properties and membrane proteins.

Tanja Grkovic is the NMR Professional Officer based at the Eskitis Institute for Drug Discovery at Griffith University in Brisbane, Australia. Her research foci include the natural products chemistry of marine microbes and natural product-based drug discovery.

Malcolm Jones is an Associate Professor of Veterinary Biology and Parasitology at the University of Queensland, Australia, and visiting scientist at the Queensland Institute of Medical Research. His research interests lie in the biology and control of pathogenic helminth (worm) infections.

Methods of
Molecular Analysis in the

CONTENTS

FOREWORD

Contemporary scientific research is in large parts an interdisciplinary effort, especially when it comes to the investigation of processes in living organisms, the so-called life sciences. It has thus become an essential requirement to have an appreciation of methodologies that neighbour one's own area of expertise. In particular areas, such as for example modern structural biology, understanding of a variety of different analytical methods that used to be the core domain of other disciplines or specialised research areas is now a mandatory requirement.

The core focus of this text is on properties of molecules and the study of their interactions. Within the life sciences, spanning diverse fields from analysis of elements in environmental or tissue samples to the design of novel drugs or vaccines, the molecules of interest thus span different orders of magnitude as well – from inorganic ions or gases as molecules with only few atoms, over small organic molecules, natural products and biomolecules, up to macromolecules such as proteins and DNA.

The methods covered in this text are featured in other textbooks, mainly in two different ways. On the one hand, many texts aimed at students contain a brief overview of particular methodologies, and mostly this is just enough to whet the appetite. On the other hand, there are authoritative in-depth treatises where the amount and level of detail in many cases exceeds the absorbing capacity of a non-expert.

The authors of this book, in contrast, have compiled a text that delivers the fundamental insights into the most popular methods of molecular analysis in a concise and accessible fashion.

This book should appeal to researchers in the area of life sciences who are not necessarily expert in all the different methodologies of molecular analysis. It should also be useful to students of chemistry and biochemistry disciplines, in particular to those studying the interactions between molecules. Teachers may find this an auxiliary text for courses in chemistry, biochemistry and biophysical chemistry, as well as forensics and environmental studies. And certainly anyone interested in the understanding of fundamental molecular analytical methods should find this text a useful and accessible introduction.

Professor Dr Robert Huber
Martinsried, 18 March 2013

PREFACE

The life sciences, comprising the study of living organisms, is the most prominent example of modern interdisciplinary research where complex processes are investigated by means of particular scientific disciplines. Important contributions are made by disciplines that study molecular structure, interactions and their implications for function.

This text is meant for everyone who studies or has an interest in molecular aspects of the life sciences. It aims to provide the background for tools and methodologies originating from the core disciplines of chemistry and physics applied to investigation of problems relevant to the life sciences.

With this text, we attempt to fill a gap by presenting relevant methodologies in a manageable volume, but with strong emphasis on describing the fundamental principles for the individual methods covered. Deliberately, we have chosen to include mathematical formulas where we found them to be of integral importance for the matter discussed. A powerful feature of mathematical equations is their ability to capture relationships between different parameters that can be complicated when described in words. Not least, almost all formulas are an essential part of the work and analysis in a scientific project and are thus a tool used in real-life applications. We hope that the combination of discussion, illustration and mathematical expressions deliver a representation of a phenomenon from different aspects, helping to form an understanding of the methodologies, rather than just a memory.

This book is in large parts based on lectures we developed at The University of Edinburgh, Griffith University, University of Lyon, and the University of Queensland. Consciously or unconsciously, many colleagues we have learned from have made contributions. Data for many figures and tables in this book have been obtained from experiments conducted particularly for this book. We are very grateful to Dr Michelle Colgrave (CSIRO, Brisbane), Dr Nien-Jen Hu (Imperial College London) and Lawren Sullivan (Griffith University) for providing experimental data used in various figures. Manuscript and figures for this book have been compiled entirely with open source and academic software under Linux, and we would like to acknowledge the efforts by software developers and programmers who make their products freely available.

Recommendations for further reading and websites of interest have been compiled based on popular acceptance as well as the authors' preferences; however, the selections evidently are not exhaustive. In cases where commercial supplier websites are listed, these have been included based purely on educational value; the authors have not received any benefit from those companies in this context.

We are particularly grateful to Professor Lindsay Sawyer (The University of Edinburgh) for many helpful suggestions and critical reading of the manuscript, and Professor Robert Huber (Max-Planck-Institute for Biochemistry, Martinsried) for his guiding advice.

Andreas Hofmann
Anne Simon
Tanja Grkovic
Malcolm Jones
March 2013

UNITS AND CONSTANTS

Decimal factors.

Factor	Prefix	Symbol	Factor	Prefix	Symbol
10^{-1}	deci	d	10	deka	da
10^{-2}	centi	c	10^{2}	hekto	h
10^{-3}	milli	m	10^{3}	kilo	k
10^{-6}	micro	μ	10^{6}	mega	M
10^{-9}	nano	n	10^{9}	giga	G
10^{-12}	pico	p	10^{12}	tera	T
10^{-15}	femto	f	10^{15}	peta	P
10^{-18}	atto	a	10^{18}	exa	E

SI base parameters and units.

Symbol	Parameter	Unit	Name
I	Electric current	A	Ampere
I	Light intensity	cd	Candela
l	Length	m	Metre
m	Mass	kg	kilogram
n	Molar amount	mol	Mol
t	Time	s	second
T	Temperature	K	Kelvin

Important physico-chemical parameters and units.

Symbol	Parameter	Unit	Name
B	Magnetic induction	$1\,T = 1\,kg\,s^{-2}\,A^{-1} = 1\,V\,s\,m^{-2}$	Tesla
c	Molar concentration	$1\,mol\,l^{-1}$	
C	Electric capacity	$1\,F = 1\,kg^{-1}\,m^{-2}\,s^{4}\,A^{2} = 1\,A\,s\,V^{-1}$	Farad
E	Energy	$1\,J = 1\,kg\,m^{2}\,s^{-2}$	Joule
ε	Molar extinction coefficient	$1\,l\,mol^{-1}\,cm^{-1}$	
F	Force	$1\,N = 1\,kg\,m\,s^{-2} = 1\,J\,m^{-1}$	Newton
Φ	Magnetic flux	$1\,Wb = 1\,kg\,m^{2}\,s^{-2}\,A^{-1} = 1\,V\,s$	Weber
G	Electric conductivity	$1\,S = 1\,kg^{-1}\,m^{-2}\,s^{3}\,A^{2} = 1\,\Omega^{-1}$	Siemens
H	Enthalpy	$1\,J = 1\,kg\,m^{2}\,s^{-2}$	Joule
L	Magnetic inductivity	$1\,H = 1\,kg\,m^{2}\,s^{-2}\,A^{-2} = 1\,V\,A^{-1}\,s$	Henry
M	Molar mass[a]	$1\,g\,mol^{-1} = 1\,Da$	(Dalton)
ν	Frequency	$1\,Hz = 1\,s^{-1}$	Hertz
p	Pressure	$1\,Pa = 1\,kg\,m^{-1}\,s^{-2} = 1\,N\,m^{-2}$	Pascal

(*cont.*)

Symbol	Parameter	Unit	Name
P	Power	$1\,W = 1\,kg\,m^2\,s^{-3} = 1\,J\,s^{-1}$	Watt
Q	Electric charge	$1\,C = 1\,A\,s$	Coulomb
ρ	Density	$1\,g\,cm^{-3}$	
ρ^*	Mass concentration	$1\,mg\,ml^{-1}$	
θ	Temperature	$1\,°C$	Celsius
R	Electric resistance	$1\,\Omega = 1\,kg\,m^2\,s^{-3}\,A^{-2} = 1\,V\,A^{-1}$	Ohm
S	Entropy	$1\,J\,K^{-1}$	
U	Electric potential (voltage)	$1\,V = 1\,kg\,m^2\,s^{-3}\,A^{-1} = 1\,J\,A^{-1}\,s^{-1}$	Volt
V	Volume	$1\,l$	
V_m	Molar volume	$1\,l\,mol^{-1}$	
v	Partial specific volume	$1\,ml\,g^{-1}$	
x	Molar ratio	1	

[a] Note that the molecular mass is the mass of one molecule given in atomic mass units. The molar mass is the mass of 1 mol of molecules and thus has the unit of $g\,mol^{-1}$.

Important physico-chemical constants.

Symbol	Constant	Value
c	Speed of light *in vacuo*	$2.99792458 \times 10^8\,m\,s^{-1}$
e	Elementary charge	$1.6021892 \times 10^{-19}\,C$
$\varepsilon_0 = (\mu_0\,c^2)^{-1}$	Electric field constant	$8.85418782 \times 10^{-12}\,A^2\,s^4\,m^{-3}\,kg^{-1}$
$F = N_A$	Faraday's constant	$9.648456 \times 10^4\,C\,mol^{-1}$
g	Earth's gravity near surface	$9.81\,m\,s^{-2}$
$g_e = 2\,\mu_e/\mu_B$	Landé factor of free electron	2.0023193134
γ_p	Gyromagnetic ratio of proton	$2.6751987 \times 10^8\,s^{-1}\,T^{-1}$
h	Planck's constant	$6.626176 \times 10^{-34}\,J\,s$
$k = k_B = R/N_A$	Boltzmann's constant	$1.380662 \times 10^{-23}\,J\,K^{-1}$
m_e	Mass of electron	$9.109534 \times 10^{-31}\,kg$
m_n	Mass of neutron	$1.6749543 \times 10^{-27}\,kg$
m_p	Mass of proton	$1.6726485 \times 10^{-27}\,kg$
μ_0	Magnetic field constant	$4\pi \times 10^{-7}\,m\,kg\,s^{-2}\,A^{-2}$
$\mu_B = eh/(4\pi m_e)$	Bohr magneton	$9.274078 \times 10^{-24}\,J\,T^{-1}$
μ_ε	Magnetic moment of electron	$9.284832 \times 10^{-24}\,J\,T^{-1}$
$\mu_N = eh/(4\pi m_p)$	Nuclear magneton	$5.050824 \times 10^{-27}\,J\,T^{-1}$
N_A, L	Avogadro's (Loschmidt's) constant	$6.022045 \times 10^{23}\,mol^{-1}$
p^0	Normal pressure	$1.01325 \times 10^5\,Pa$
R	Gas constant	$8.31441\,J\,K^{-1}\,mol^{-1}$
R_∞	Rydberg's constant	$1.097373177 \times 10^7\,m^{-1}$
θ_0	Zero at Celsius scale	273.15 K
$v^0 = RT^0/p^0$	Molar volume of an ideal gas	$22.413831\,mol^{-1}$

Conversion factors for energy.

	1 J	1 cal	1 eV
1 J	1	0.2390	$6.24150974 \times 10^{18}$
1 cal	4.184	1	2.612×10^{19}
1 eV	$1.60217646 \times 10^{-19}$	3.829×10^{-20}	1

Conversion factors for pressure.

	1 Pa	1 atm	1 mm Hg (Torr)	1 bar
1 Pa	1	9.869×10^{-6}	7.501×10^{-3}	10^{-5}
1 atm	1.013×10^{5}	1	760.0	1.013
1 mm Hg (Torr)	133.3	1.316×10^{-3}	1	1.333×10^{-3}
1 bar	10^{5}	0.9869	750.1	1

Molar masses of amino acids, free and within peptides.

Amino acid			M (g mol^{-1})	M – M(H_2O) (g mol^{-1})
A	Ala	Alanine	89	71
C	Cys	Cysteine	121	103
D	Asp	Aspartic acid	133	115
E	Glu	Glutamic acid	147	129
F	Phe	Phenylalanine	165	147
G	Gly	Glycine	75	57
H	His	Histidine	155	137
I	Ile	Isoleucine	131	113
K	Lys	Lysine	146	128
L	Leu	Leucine	131	113
M	Met	Methionine	149	131
N	Asn	Asparagine	132	114
P	Pro	Proline	115	97
Q	Gln	Glutamine	146	128
R	Arg	Arginine	174	156
S	Ser	Serine	105	87
T	Thr	Threonine	119	101
V	Val	Valine	117	99
W	Trp	Tryptophan	204	186
Y	Tyr	Tyrosine	181	163

Introduction

1

1.1 Electromagnetic radiation

Light is a form of electromagnetic radiation, usually a mixture of waves having different wavelengths. Spectroscopic applications in structural laboratories are concerned with light from different wavelength intervals. Figure 1.1 presents an overview of different spectroscopic techniques and the energy intervals they operate in.

Many spectroscopic techniques in structural biology use light within the range of visible (Vis) colours extended on each side of the spectrum by the ultraviolet (UV) and the infrared (IR) regions (Table 1.1); these techniques are usually called spectrophotometric techniques.

1.1.1 Properties of electromagnetic radiation

The interaction of electromagnetic radiation with matter is a quantum phenomenon and dependent upon both the properties of the radiation and the appropriate structural parts of the samples involved. This is not surprising, as

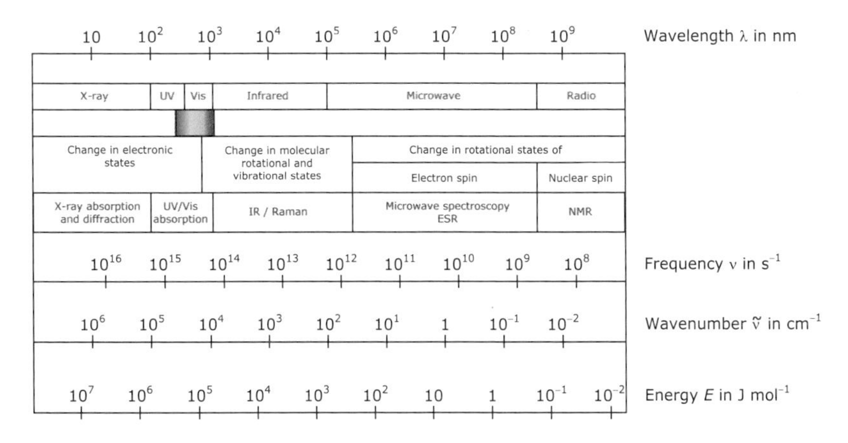

Fig. 1.1. The electromagnetic spectrum and its usage for spectroscopic methods.

Table 1.1. The three common light types for spectrophotometry.

	Wavelength (nm)	Wavenumber (cm^{-1})	Frequency (Hz)	Energy (eV)
UV	100–400	100 000–25 000	2.99×10^{15}–7.50×10^{14}	12.4–3.1
Vis	400–700	25 000–14 286	7.50×10^{14}–4.28×10^{14}	3.1–1.8
IR	700–15 000	14 286–667	4.28×10^{14}–2.00×10^{13}	1.8–0.08

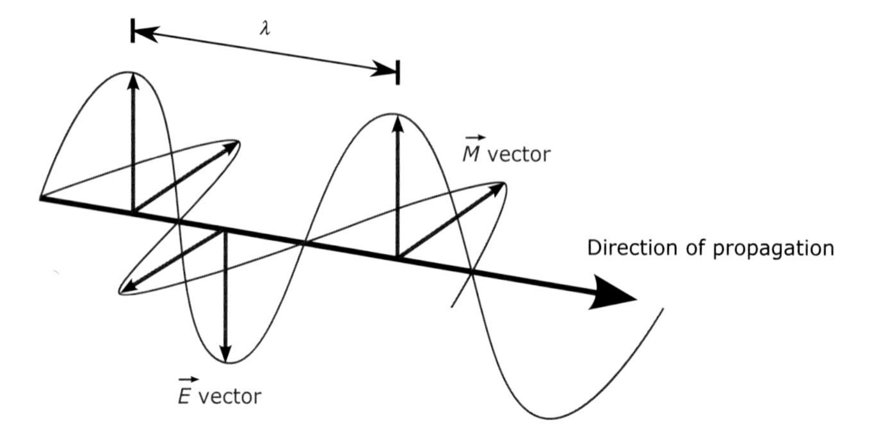

Fig. 1.2. Light is electromagnetic radiation and can be described as a wave propagating in space and time. The electric ($\vec{E}$) and magnetic ($\vec{M}$) field vectors are directed perpendicular to each other. For UV/Vis, circular dichroism and fluorescence spectroscopy, the electric field vector is of more importance. For electron paramagnetic and nuclear magnetic resonance, the emphasis is on the magnetic field vector.

the origin of electromagnetic radiation is due to energy changes within matter itself. The transitions that occur within matter are quantum phenomena, and the spectra that arise from such transitions are predictable in principle.

Electromagnetic radiation (Fig. 1.2) is composed of an electric vector ($\vec{E}$) and a perpendicular magnetic vector ($\vec{M}$), each oscillating in a plane at right angles to the direction of propagation. The wavelength λ is the spatial distance between two consecutive peaks (one cycle) in the sinusoidal waveform and is measured in multiples of nanometres (nm). The maximum length of the vector is called the amplitude. The frequency ν of the electromagnetic radiation is the number of oscillations made by the wave within the time frame of 1 s. It therefore has the unit of $1\ s^{-1} = 1$ Hz. The frequency is related to the wavelength via the speed of light $c = 2.998 \times 10^8\ m\ s^{-1}$ (*in vacuo*) by $\nu = c\lambda^{-1}$. A related parameter in this context is the wavenumber

$$\tilde{\nu} = \frac{1}{\lambda}, \tag{1.1}$$

which describes the number of completed wave cycles per distance and is typically measured as cm^{-1}.

1.1.2 Interaction of light with matter

Figure 1.1 shows the spectrum of electromagnetic radiation organised by increasing wavelength, and thus decreasing energy, from left to right. Also annotated are the types of radiation, the various interactions with matter and the resulting spectroscopic applications, as well as the interdependent parameters of frequency and wavenumber.

Electromagnetic phenomena are explained in terms of quantum mechanics. The photon is the elementary particle responsible for electromagnetic phenomena. It carries the electromagnetic radiation and has properties of a wave, as well as of a particle, albeit having a mass of zero. As a particle, it interacts with matter by transferring its energy E:

$$E = \frac{\mathrm{hc}}{\lambda} = \mathrm{h}\nu, \tag{1.2}$$

where h is the Planck constant ($\mathrm{h} = 6.63 \times 10^{-34}$ J s) and ν is the frequency of the radiation as introduced above.

When considering a diatomic molecule (see Fig. 1.3), rotational and vibrational levels possess discrete energies that only merge into a continuum at very high energy. Each electronic state of a molecule possesses its own set of rotational and vibrational levels. As the kind of schematics shown in Fig. 1.3 is rather complex, the Jablonski diagram is used instead, where electronic and vibrational states are schematically drawn as horizontal lines, and vertical lines depict possible transitions (see Figs 1.5 and 2.14).

In order for a transition to occur in the system, energy must be absorbed. The energy change ΔE needed for the transition is defined in quantum terms by the difference in absolute energies between the final and the starting state as $\Delta E = E_{\mathrm{final}} - E_{\mathrm{start}} = \mathrm{h}\nu$.

Electrons in either atoms or molecules may be distributed between several energy levels but principally reside in the lowest levels (ground state). In order for an electron to be promoted to a higher level (excited state), energy must be put into the system. If this energy $E = \mathrm{h}\nu$ is derived from electromagnetic radiation, this gives rise to an absorption spectrum, and an electron is transferred from the electronic ground state (S_0) into the first electronic excited state (S_1). Note that this requires an exact match of the photon energy with the energy difference between the two states that the transition is occurring between (resonance condition). The molecule will also be in an excited vibrational and rotational state. Subsequent relaxation of the molecule into the vibrational ground state of the first electronic excited state will occur. The electron can then revert back to the electronic ground state. For non-fluorescent molecules, this is accompanied by the emission of heat (ΔH).

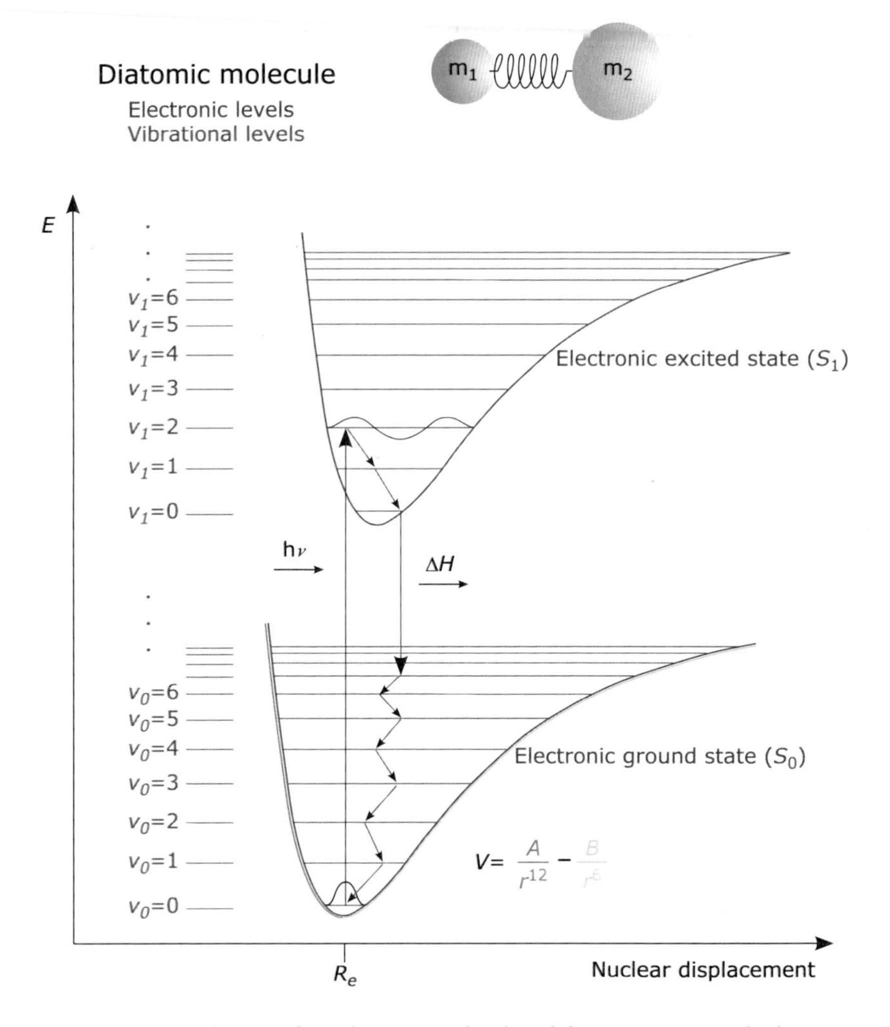

Fig. 1.3. Energy diagram for a diatomic molecule exhibiting rotation and vibration as well as an electronic structure. The energy of a system as a function of distance between two masses, m_1 and m_2 (nuclear displacement), is described as a Lennard–Jones potential curve (blue) with different equilibrium distances (R_e) for each electronic state. Energetically lower states always have lower equilibrium distances. The vibrational levels (quantum number v; red) are superimposed on the electronic levels. Rotational levels are superimposed on each vibrational level and not shown for reasons of clarity. The Lennard–Jones potential V is the most widely used model and consists of the van der Waals attractive term ($\sim r^{-6}$; green) and the Pauli repulsive term ($\sim r^{-12}$; red); A and B are constants.

Chromophores are the light-absorbing moieties within a molecule. Due to differences in electronegativity between individual atoms, they possess a spatial distribution of electric charge. This results in a dipole moment $\vec{\mu}_0$, such as for example the permanent dipole moment of the peptide bond

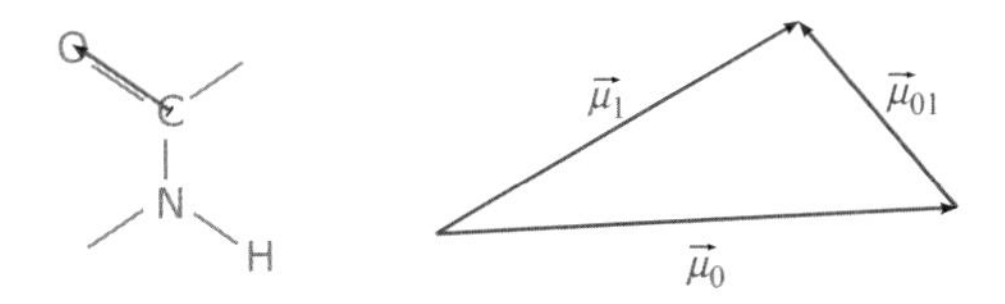

Fig. 1.4. Left: Dipole moment of the peptide bond. Right: The transition dipole moment $\vec{\mu}_{01}$ is the difference vector between the dipole moment of the chromophore in the ground state $\vec{\mu}_0$ and the excited state $\vec{\mu}_1$.

(Fig. 1.4). When light is absorbed by the chromophore, the distribution of electric charge is altered and the dipole moment changes accordingly ($\vec{\mu}_1$). The transition dipole moment $\vec{\mu}_{01}$ is the vector difference between the dipole moment of the chromophore in the ground and the excited state (Fig. 1.4). The transition dipole moment is a measure for transition probability. The dipole strength of the transition dipole moment, D_{01}, is defined as the squared length of the transition dipole moment vector:

$$D_{01} = |\overrightarrow{\mu_{01}}|^2 \tag{1.3}$$

Transitions with $D_{01} \rightarrow 0$ are called forbidden transitions and the probability of their occurrence is low. If $D_{01} \rightarrow 1$, the transition is said to be 'allowed' and occurs with high probability.

The plot of absorption probability against wavelength is called the absorption spectrum. In the simpler case of single atoms (as opposed to multi-atom molecules), electronic transitions lead to the occurrence of line spectra (see Section 2.1). Because of the existence of vibrational and rotational energy levels in the different electronic states, molecular spectra are usually observed as band spectra (for example Fig. 1.5), which are molecule specific due to the unique vibration states.

A commonly used classification of absorption transitions uses the spin states of electrons. Quantum mechanically, the electronic states of atoms and molecules are described by orbitals, which define the different states of electrons by two parameters: a geometrical function defining the space and a probability function. The combination of both functions describes the localisation of an electron.

In systems comprising more than one atom (molecules), the individual atomic orbitals combine into molecular orbitals (linear combination of atomic orbitals (LCAO); see Fig. 1.6).

Electrons in bonding orbitals are usually paired with anti-parallel spin orientation (Fig. 1.7). The total spin S is calculated as the sum of the individual electron spins. The multiplicity M is obtained by

$$M = 2S + 1. \tag{1.4}$$

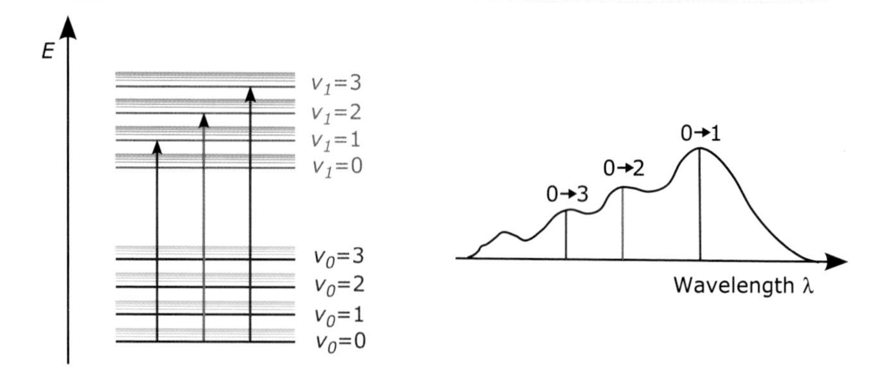

Fig. 1.5. Transitions in polyatomic molecules give rise to band spectra (as opposed to the line spectra observed with single atoms). Left: Vibrational (black, grey) and rotational (green) energy levels exist in the electronic ground state and excited state. Absorption of UV/Vis light leads to a general transition from the electronic ground to the excited state, but different vibrational and rotational levels are populated according to quantum mechanical rules. Right: As the lines for particular vibrational and rotational transitions are energetically close, they cannot be resolved individually and lead to the appearance of 'bands'. The main peaks represent the vibrational transitions. The rotational transitions are typically not resolved in UV/Vis absorption spectra.

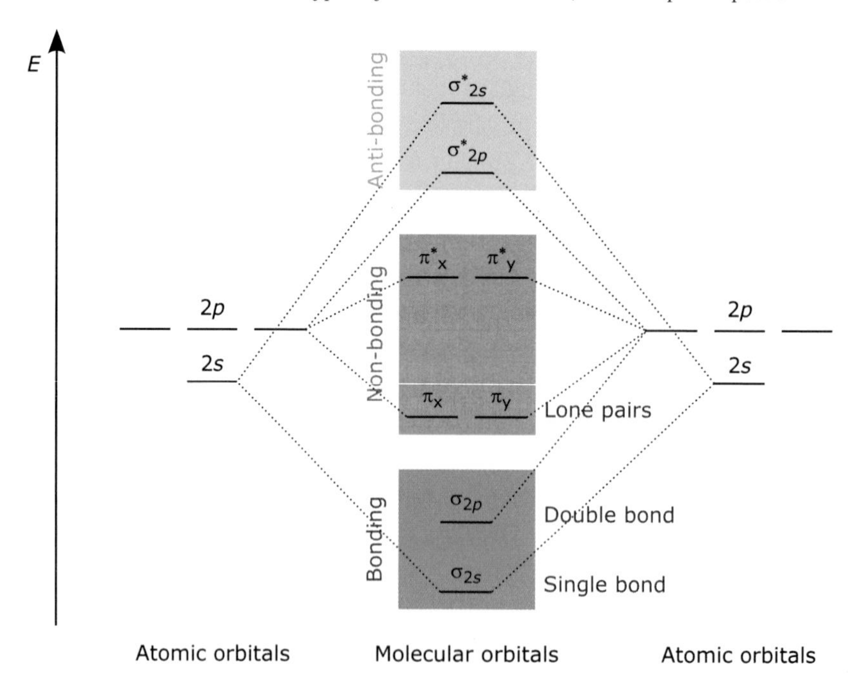

Fig. 1.6. Illustration of the linear combination of atomic orbitals (LCAO) to form molecular orbitals. Two atoms combine their 2*s* and 2*p* orbitals to form the molecular orbitals shown in the centre. The orbitals occupied by electrons responsible for single bond, double bond and lone electron pairs are indicated. Occupation of the anti-bonding orbitals leads to destabilisation of the bonds (bond cleavage).

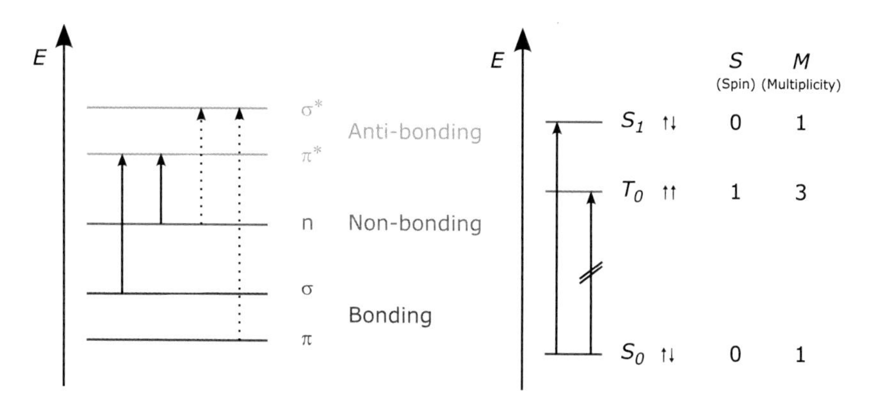

Fig. 1.7. Left: Energy scheme for molecular orbitals (not to scale). Arrows indicate possible electronic transitions. The length of the arrows indicates the energy required to be put into the system in order to enable the transition. Solid arrows depict transitions possible with energies from the UV/Vis spectrum for some biological molecules. The transitions shown by dotted arrows require higher energies (e.g. X-rays). Right: Due to quantum mechanical rules, only transitions with conserved total spin S are allowed. The transition between the two singlet states $S_0 \rightarrow S_1$ is therefore allowed; the transition from singlet ground state S_0 to the triplet state T_1 is forbidden.

For paired electrons in one orbital, this yields:

$$S = \text{spin}_{\text{electron1}} + \text{spin}_{\text{electron2}} = \left(+\frac{1}{2}\right) + \left(-\frac{1}{2}\right) = 0. \tag{1.5}$$

The multiplicity is thus:

$$M = 2 \times 0 + 1 = 1. \tag{1.6}$$

Such a state is thus called a singlet state and denoted 'S' (not to be confused with the total spin S above). Usually, the ground state of a molecule is a singlet state, S_0.

In cases where the spins of both electrons are oriented in a parallel fashion, the resulting state is characterised by a total spin of $S = 1$, and a multiplicity of $M = 3$. Such a state is called a triplet state and usually exists only as one of the excited states of a molecule, e.g. T_1.

According to quantum mechanical transition rules, the multiplicity M and the total spin S must not change during a transition. Thus, the $S_0 \rightarrow S_1$ transition is allowed and possesses a high transition probability. In contrast, the $S_0 \rightarrow T_1$ transition is not allowed and has a small transition probability. Note that the intensity of an absorption band is proportional to the transition probability.

Most biologically relevant molecules possess more than two atoms and therefore the energy diagrams become more complex than the ones shown in Fig. 1.3. Different orbitals combine to yield molecular orbitals that generally fall into one of five different classes (Fig. 1.7): *s* orbitals combine to the bonding σ and the anti-bonding σ^* orbitals. Some *p* orbitals combine to the bonding π and the anti-bonding π^* orbitals. Other *p* orbitals combine to form non-bonding *n* orbitals. The population of bonding orbitals strengthens a chemical bond, while the population of anti-bonding orbitals weakens a chemical bond.

1.2 Lasers

Laser is an acronym for *l*ight *a*mplification by *s*timulated *e*mission of *r*adiation. A detailed explanation of the theory of lasers is beyond the scope of this textbook. A simplified description starts with the use of photons of a defined energy to excite an absorbing material. This results in elevation of an electron to a higher energy level. If, while the electron is in the excited state, another photon of precisely that energy arrives, then, instead of the electron being promoted to an even higher level, it can return to the original ground state. However, this transition is accompanied by the emission of two photons with the same wavelength and exactly in phase (coherent photons). Multiplication of this process will produce coherent light with an extremely narrow spectral bandwidth. In order to produce an ample supply of suitable photons, the absorbing material is surrounded by a rapidly flashing light of high intensity ('pumping').

Lasers are indispensable tools in many areas of science, including biochemistry and biophysics. Several modern spectroscopic techniques utilise laser light sources, due to their high intensity and accurately defined spectral properties. One of the probably most revolutionising applications in the life sciences, the use of lasers in DNA sequencing with fluorescent labels (see Section 2.3.6), enabled the breakthrough in whole-genome sequencing.

FURTHER READING

Banwell, C. N. & McCash, E. M. (1994). *Fundamentals of Molecular Spectroscopy*, 4th edn. London: McGraw-Hill. (A readable account of spectroscopic principles.)

Cantor, C. R. & Schimmel, P. R. (1980). *Biophysical Chemistry*, 1st edn. New York: Freeman. (A comprehensive reference in three parts.)

Hoppe, W., Lohmann, W., Markl, H. & Ziegler, H. (1982). *Biophysik*, 2nd edn. Berlin, Heidelberg, New York: Springer Verlag. (A rich and authoritative compendium of the physical basics of the life sciences.)

Websites

Physics 2000: an interactive journey through modern physics, including topics such as waves, quantum mechanics, etc.: http://www.colorado.edu/physics/2000/

Applet: Spectrum: http://lectureonline.cl.msu.edu/%7Emmp/applist/Spectrum/s.htm

2 Spectroscopic methods

2.1 Atomic spectroscopy

Electronic transitions in single atoms yield clearly defined line spectra. In atomic emission spectroscopy (AES), these lines can be observed as light of a particular wavelength (colour). Conversely, black lines can be observed against a bright background in atomic absorption spectroscopy (AAS). The wavelengths emitted from excited atoms may be identified using a spectroscope with the human eye as the 'detector', or with a spectrophotometer. Both methods are used mainly for the analysis of metals, as well as some metalloids (such as boron, silicon, germanium, arsenic, antimony and tellurium) and a few non-metals (phosphorus).

Using light of high energy (e.g. X-rays), fluorescence from an electronic transition within individual atoms can be elicited.

2.1.1 Principles

In a spectrum of an element, the absorption or emission wavelengths are associated with transitions that require a minimum of energy change. In order for energy changes to be minimal, transitions tend to occur between orbitals close together in energy terms. For example, excitation of a sodium atom and its subsequent relaxation gives rise to emission of orange light ('D-line') due to the transition of an electron between the 3*s* and 3*p* orbitals (Fig. 2.1).

Electron transitions in an atom are limited by the availability of empty orbitals. Filling orbitals with electrons is subject to three major rules:

- Any one orbital can be occupied by a maximum of two electrons.
- The spins of electrons in one orbital need to be paired in an anti-parallel fashion (Pauli principle).
- Energetically equivalent orbitals are first to be filled with electrons of parallel spin; only then can a second electron with anti-parallel spin be added (Hund's rule).

Taken together, these limitations mean that emission and absorption lines are characteristic for an individual element.

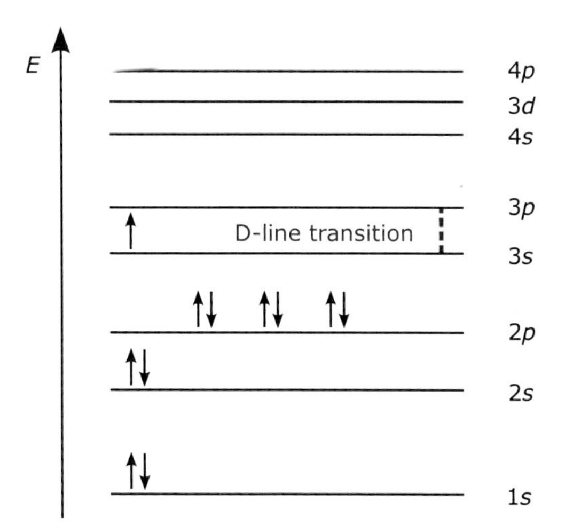

Fig. 2.1. Atomic orbitals of Na. The transition of the electron in the 3*s* orbital into the 3*p* orbital gives rise to the D-line in the atomic spectrum of Na. The elevation into the higher state is seen as a line in the absorption spectrum, the return into the lower state is observed as a line in the emission spectrum.

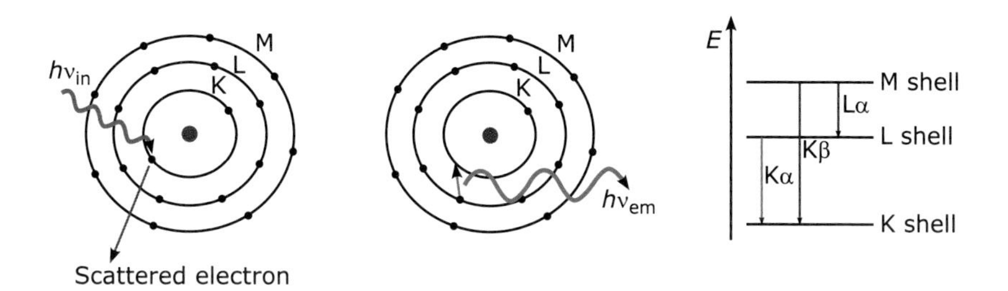

Fig. 2.2. The principle of X-ray fluorescence illustrated with Bohr's atom model (Bohr, 1913a, b, c). Left: An incoming X-ray photon leads to ejection of a core electron from the atom. Middle: The vacancy is filled with an electron from a higher shell. This results in emission of a photon whose energy is equal to the difference in the energies of the two shells (here: L → K). Right: Energy scheme illustrating the main transitions observed in atomic fluorescence spectroscopy.

If light of higher energy is provided, such as for example X-rays, ionisation of an atom may occur (Fig. 2.2). In this case, an electron is ejected from a core shell (typically a K- or L-shell) of the atom, leading to an unstable electronic structure with a missing negative charge close to the positively charged atom nucleus. An electron from an outer shell (i.e. an energetically higher orbital) transitions to take the place of the missing electron of the core shell. This transition is accompanied by emission of a photon with an energy equivalent to the energy gap between the two orbitals.

This process is called atomic fluorescence. The main transitions observed are:

- L → K (Kα transition)
- M → K (Kβ transition)
- M → L (Lα transition)

Due to characteristic orbital energies of each element, the energy difference (equal to the energy of the emitted photon) can be used for identification of elements in samples. Theoretically, the lightest element to be determined by this method is beryllium; however, as very light elements possess low atomic fluorescence yields, experimentally, only elements up to sodium can be determined. Atomic absorption and emission spectroscopy are essential tools for elementary analysis from star light. Another application for the emitted photons is in generation of X-ray light (see Fig. 3.16).

2.1.2 Instrumentation

In general, atomic spectroscopy is not carried out in solution. In order for atoms to emit or absorb monochromatic radiation, they need to be volatilised by exposing them to high thermal energy. Usually, nebulisers are used to spray the sample solution into a flame or an oven. Alternatively, the gaseous form can be generated by using an inductively coupled plasma (ICP). The variation in temperature and composition of a flame make standard conditions difficult to achieve. Most modern instruments thus use an ICP.

Atomic emission spectroscopy and AAS are generally used to identify specific elements present in the sample and to determine their concentrations. The energy absorbed or emitted is proportional to the number of atoms in the optical path. Strictly speaking, in the case of emission, it is the number of excited atoms that is proportional to the emitted energy. Concentration determination with AES or AAS is carried out by comparison with calibration standards.

2.1.3 Experimental procedures

Sodium gives rise to high background readings and is usually measured first. A similar amount of sodium is then added to all other standards. Excess hydrochloric acid is commonly added, because chloride compounds are often the most volatile salts. Calcium and magnesium emission can be enhanced by the addition of alkali metals and suppressed by the addition of phosphate, silicate and aluminate, as these form non-dissociable salts. The suppression effect can be relieved by the addition of lanthanum and strontium salts. Lithium is frequently used as an internal standard. For storage of samples and standards, polyethylene bottles are used, as glass can absorb and release metal ions and thus impact the accuracy of this sensitive technique.

Cyclic analysis may be performed that involves the estimation of each interfering substance in a mixture. Subsequently, the standards for each component in the mixture are doped with each interfering substance. This process is repeated two or three times with refined estimates of interfering substance, until self-consistent values are obtained for each component.

Flame instability requires experimental protocols where determination of an unknown sample is bracketed by measurements of the appropriate standard, in order to achieve the highest possible accuracy.

Biological samples are usually converted to ash prior to determination of metals. Wet ashing in solution is often used, employing an oxidative digestion similar to the Kjeldahl method.

2.1.4 Applications

Atomic emission and atomic absorption spectrophotometry

Sodium and potassium are assayed at concentrations of a few parts per million (ppm; i.e. in the order of mg l^{-1}) using simple filter photometers. The modern emission spectrophotometers allow determination of about 20 elements in biological samples, the most common being calcium, magnesium and manganese. Absorption spectrophotometers are usually more sensitive than emission instruments and can detect less than 1 ppm of each of the common elements with the exception of alkali metals. The relative precision is about 1% in a working range of 20–200 times the detection limit of an element.

Atomic emission spectroscopy and AAS have been widely used in analytical chemistry, such as environmental and clinical laboratories. Nowadays, the technique has been superseded largely by the use of ion-selective electrodes.

Atomic fluorescence spectrophotometry

Atomic fluorescence spectrophotometry is a technique applied in in-house laboratories and uses the same basic setup as AES and AAS. The atoms need to be vapourised by one of three methods (flame, electric or

ICP). Once vapourised, the atoms are excited using electromagnetic radiation by directing a light beam into the sample. This beam must be intense, but not spectrally pure, as only the resonant wavelengths will be absorbed, leading to fluorescence (see Section 2.3). Despite being limited to only a few metals, the main importance of this technique lies in the extreme sensitivity. For example, zinc and cadmium can be detected at levels as low as 10–20 parts per trillion (ppt; 1×10^{12}, i.e. in the order of ng l^{-1}).

2.1.5 X-ray fluorescence microscopy

The use of synchrotron X-ray light to excite atoms to stimulate fluorescence emission provides a micro-analytical tool for mapping of elemental distribution in samples. By focusing a narrow primary X-ray beam into the plane of a confocal light microscope as well as using energy-dispersion detectors with high spatial resolution and short processing times (e.g. silicon drift detectors), this methodology allows X-ray fluorescence microscopy.

Samples are mounted in the focal planes of the primary X-ray beam and the light microscope. For measurements, the sample holder is then repositioned using high-precision stepping motors so that the incident X-ray beam excites spatially different areas of the sample (rasterisation). For each raster point, fluorescence data are acquired. The data can be sorted by energy (energy-dispersive analysis) or wavelength (wavelength-dispersive analysis), and individual elements can be identified based on their characteristic spectra, provided there is no overlap between the spectra of the present elements. Typically, the data are fitted for the presence of multiple elements (global fit).

X-ray fluorescence microscopy is the ideal tool to study the intracellular distribution of metals in biochemical, diagnostic, forensic or archaeological applications to determine the presence of trace elements, toxic heavy metals and therapeutic or diagnostic metal complexes. It is also used as a process control tool in many extractive and processing industries. Compared with other micro-analytical methods for elemental distributions, X-ray fluorescence microscopy is the only technique that allows non-destructive investigation of hydrated samples (e.g. biological samples such as whole cells or tissue sections; see Fig. 2.3) with extremely high sensitivity (e.g. trace metal detection) and submicron spatial resolution. In many cases, the oxidation or coordination state of the metal can also be determined.

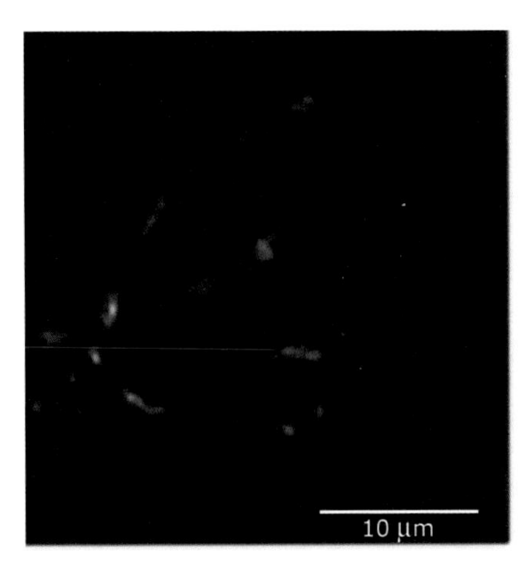

Fig. 2.3. X-ray fluorescence micrograph of a *Schistosoma japonicum* whole egg showing localisation of iron in the egg. The data shown were recorded on the channel for the Fe Kα transition ($E \approx 6400$ eV, i.e. $\lambda \approx 1.94$ Å).

2.2 UV/Vis spectroscopy

2.2.1 Light absorption

The energy of an incoming photon is used to promote electrons from the electronic ground state (S_0) to an electronic excited state (e.g. S_1). This is called an electronic transition and is caused by absorption of the incoming photon. Molecular (sub)structures responsible for interaction with electromagnetic radiation are called chromophores. The particular frequencies at which light is absorbed depend on the structure and environment of the absorbing molecule. Superimposed on electronic states are vibrational as well as rotational states of the molecule. While electronic and vibrational states are within the energy interval of UV/Vis light, the rotational states are excited by light of much lower energy (higher wavelengths), i.e. microwaves. Excited electrons return to the vibrational ground state of the electronic excited state ($v_1 = 0$) by vibrational transitions in small energy decrements. Ultimately, the electron returns to the electronic ground state. The energy released upon this relaxation process heats the environment of the molecule and is taken up by the solution (see Fig. 1.3).

The presence of rotational and vibrational states in molecules gives rise to band spectra, as opposed to the line spectra observed with single atoms (see Section 1.1.2). An important parameter of molecular spectra is therefore the wavelength at which the highest absorption of light occurs; this wavelength is called λ_{max}.

The UV/Vis region of the electromagnetic spectrum and the associated techniques are probably the most widely used for analytical work and research into biological problems.

The electronic transitions in molecules can be classified according to the participating molecular orbitals (see Fig. 1.7). From the four possible transitions ($n \rightarrow \pi^*$, $\pi \rightarrow \pi^*$, $n \rightarrow \sigma^*$ and $\sigma \rightarrow \sigma^*$), only two can be elicited with light from the UV/Vis spectrum for some biological molecules: $n \rightarrow \pi^*$ and $\pi \rightarrow \pi^*$. The $n \rightarrow \sigma^*$ and $\sigma \rightarrow \sigma^*$ transitions are energetically not within the range of UV/Vis spectroscopy and require higher energies.

A closer examination of the molecular orbitals involved in the $n \rightarrow \pi^*$ and $\pi \rightarrow \pi^*$ transitions reveals that the n or π orbitals represent the highest occupied molecular orbitals (HOMOs) while the π^* orbitals represent the lowest unoccupied molecular orbitals (LUMOs). UV/Vis absorption bands are therefore an important source of experimental data for quantum mechanics, as they enable determination of HOMO–LUMO gap energies.

A very important biological application of UV/Vis chromophores comprises photosynthetic processes, which thrive on the harvest of energy from UV/Vis light and the fast transfer of electrons from the light-harvesting regions of plants into regions of chemical reaction and synthesis.

In proteins, there are three types of chromophores relevant for UV/Vis spectroscopy:

- peptide bonds (amide bond);
- certain amino acid side chains (mainly tryptophan and tyrosine); and
- certain prosthetic groups and coenzymes (e.g. porphyrin groups such as in haem).

Molecules (chromophores) that strongly absorb light in the visual part of the electromagnetic spectrum are also called pigments, as they can readily be identified by the human eye.

The presence of several conjugated double bonds in organic molecules results in an extended π system of electrons, which lowers the energy of the π^* orbital through electron delocalisation. In many cases, such systems possess $\pi \rightarrow \pi^*$ transitions in the UV/Vis region of the electromagnetic spectrum. Such molecules include the bases of nucleic acids, prosthetic groups and coenzymes, as well as generally organic molecules with polyene groups. These types of chromophore are very useful tools in colorimetric applications (see Table 2.9).

Many inorganic anions also possess UV/Vis chromophores, but a major group of inorganic compounds amenable to UV/Vis spectroscopy is the group of charge-transfer complexes (see Section 2.2.7) which are frequently employed for practical analytical assays.

2.2.2 Intrinsic chromophores of proteins

The most useful chromophores for protein/peptide samples with suitable absorption bands in the UV/Vis wavelength range are the peptide bond, amino acid side chains (Fig. 2.4) and prosthetic groups.

The electronic transitions of the peptide bond occur in the far-UV. The intense peak at 190 nm, and the weaker one at ~210–220 nm is due to the $\pi \rightarrow \pi^*$ and $n \rightarrow \pi^*$ transitions. A number of amino acids (Asp, Glu, Asn, Gln, Arg and His) have weak electronic transitions at around 210 nm. Usually, these cannot be observed in proteins because they are masked by the more intense peptide bond absorption. The most useful range for proteins is above 230 nm, where there are absorptions from aromatic side chains (Table 2.1). While a very weak absorption maximum of phenylalanine occurs at 257 nm, tyrosine and tryptophan dominate the typical protein spectrum with their absorption maxima at 274 nm and 280 nm,

Table 2.1. Peak wavelengths and molar absorption coefficients of the main intrinsic protein chromophores in water in the UV/Vis region.

Chromophore	λ_{max} (nm)	ε (l mol^{-1} cm^{-1})
Peptide bond	~222	9 000
	~190	50 000
Trp	280	5 600
Tyr	274	1 400
Phe	257	200
His	211	5 900
Cys_2	250	300

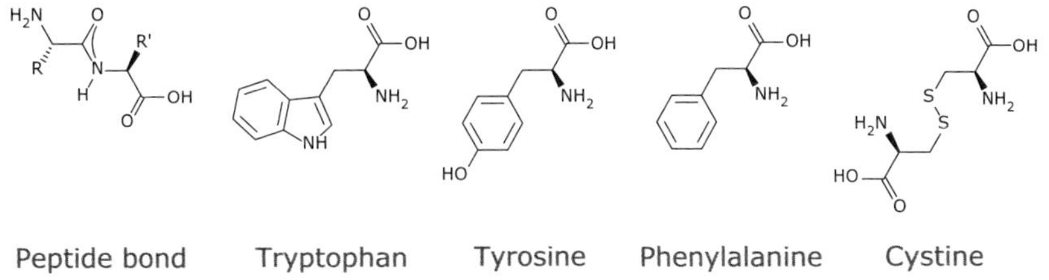

Peptide bond   Tryptophan   Tyrosine   Phenylalanine   Cystine

Fig. 2.4. Protein chromophores: The peptide bond, as well as tryptophan and tyrosine side chains, are the most important contributors to UV/Vis absorption of proteins. UV/Vis absorbance by phyenylalanine and cystine side chains is low.

respectively (see Fig. 2.13). In practice, the presence of these two aromatic side chains gives rise to a band at ~278 nm. Cystine (Cys_2) possesses a weak absorption maximum at 250 nm of similar strength to that of phenylalanine. This band can play a role in rare cases in protein optical activity or protein fluorescence.

2.2.3 Extrinsic chromophores of proteins: prosthetic groups and coenzymes

Porphyrins are the prosthetic groups of haemoglobin, myoglobin, catalase and cytochromes (Fig. 2.5). They are also the essential component of chlorophyll, which is a group of molecules of pivotal importance for light-harvesting organisms. Electron delocalisation extends throughout the cyclic tetrapyrrole ring of porphyrins and gives rise to an intense $\pi \rightarrow \pi^*$ transition at ~400 nm called the Soret band. The spectrum of haemoglobin is very sensitive to changes in the iron-bound ligand. These changes can be used for structure–function studies of haem proteins.

Molecules such as the isoalloxazine-derivative $FADH_2$, FAD (flavin adenine dinucleotide), NADH and NAD^+ (nicotinamide adenine dinucleotide; Fig. 2.6) are important coenzymes of proteins involved in electron transfer reactions (RedOx reactions). They can be conveniently assayed by using their UV/Vis absorption: 450 nm (FAD), 339 nm (NADH) and 259 nm (NAD^+) (Table 2.2). Many different biochemical RedOx reactions can be monitored by using coupling with the NAD^+/NADH RedOx pair (Fig. 2.7).

Other important chromophores in this context are carotenoids, a large class of plant pigments with red, orange and yellow colours. These molecules are composed of covalently linked isoprene units and thus constitute extended conjugated double-bond systems. Their UV/Vis spectrum is characterised by three maxima in the visible region. Notably, the stereo-configuration of such polyene systems affects their UV/Vis spectra, as can readily be seen from the different spectral characteristics of β-carotene *cis*/*trans* isomers (Fig. 2.8).

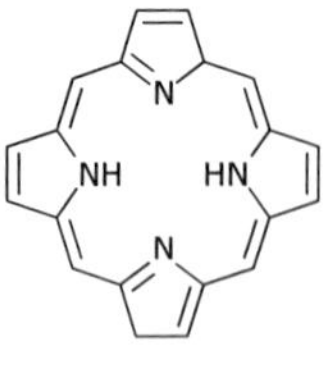

N
NH HN
N
Porphyrin

Fig. 2.5. Porphyrin is a prosthetic group of proteins and implicated in electron transfer reactions and metal coordination.

Table 2.2. Peak wavelengths (λ_{max}) and molar absorption coefficients (ε) of selected coenzymes in the UV/Vis region.

Chromophore	λ_{max} (nm)	ε (l mol^{-1} cm^{-1})
all-trans-β-carotene[a]	426	105 300
	451	139 170
	478	112 310
9-*cis*-β-carotene[a]	422	82 990
	451	123 850
	475	106 230
15-*cis*-β-carotene[a]	429	60 000
	450	84 270
	478	68 690
NADH[b]	339	6 220
NAD^{+}[b]	259	16 900
$FADH_2$[b]	No peaks	
FAD[b]	450	11 300
	473	9 200
FMN[b]	473	9 200

[a] Measured in hexane.
[b] Measured in aqueous solution.

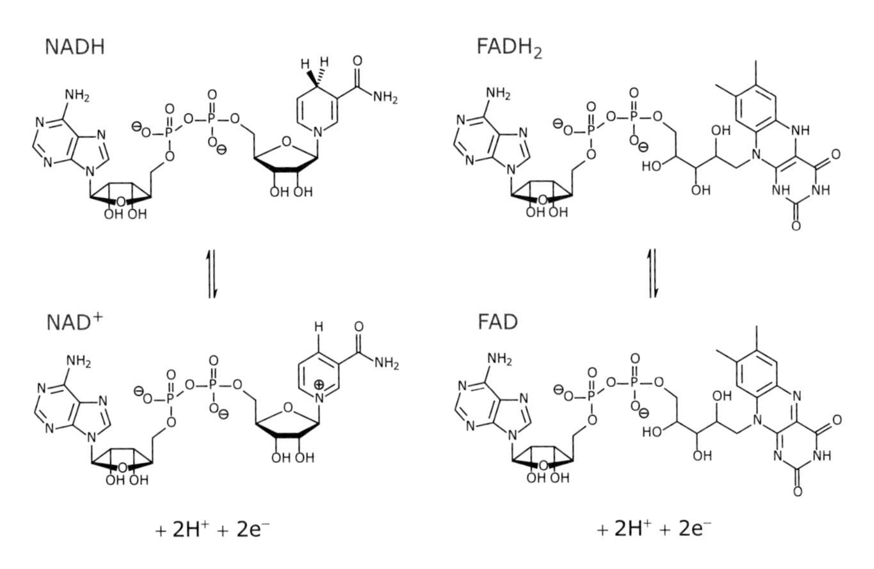

Fig. 2.6. Examples for coenzyme chromophores. RedOx reactions of nicotinamide adenine dinucleotide ($NADH/NAD^{+}$, left) and flavine adenine dinucleotide ($FADH_2$/FAD, right).

Carotenes are part of the photosynthetic machinery and are thus found in all plants. In the human intestine, carotenes are broken down into retinal, a form of vitamin A. Retinal itself is a key molecule in vision processes in the human eye and binds to the G-protein-coupled receptor opsin.

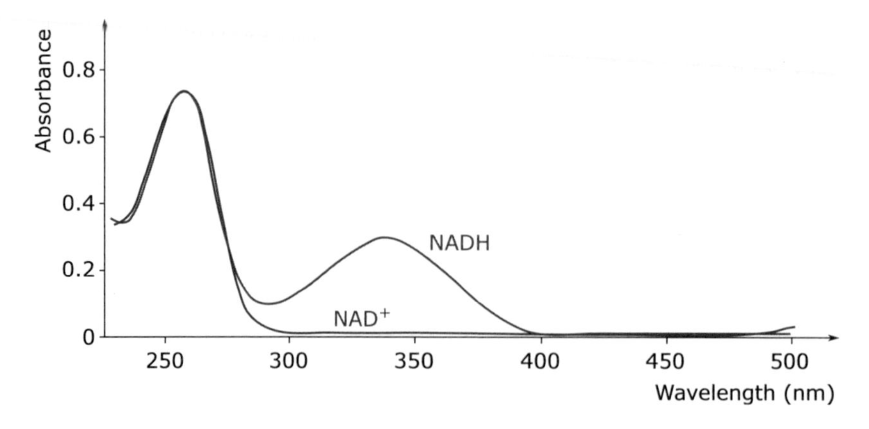

Fig. 2.7. Comparison of UV/Vis spectra of the oxidised (NAD^+) and reduced (NADH) forms of nicotinamide adenine dinucleotide.

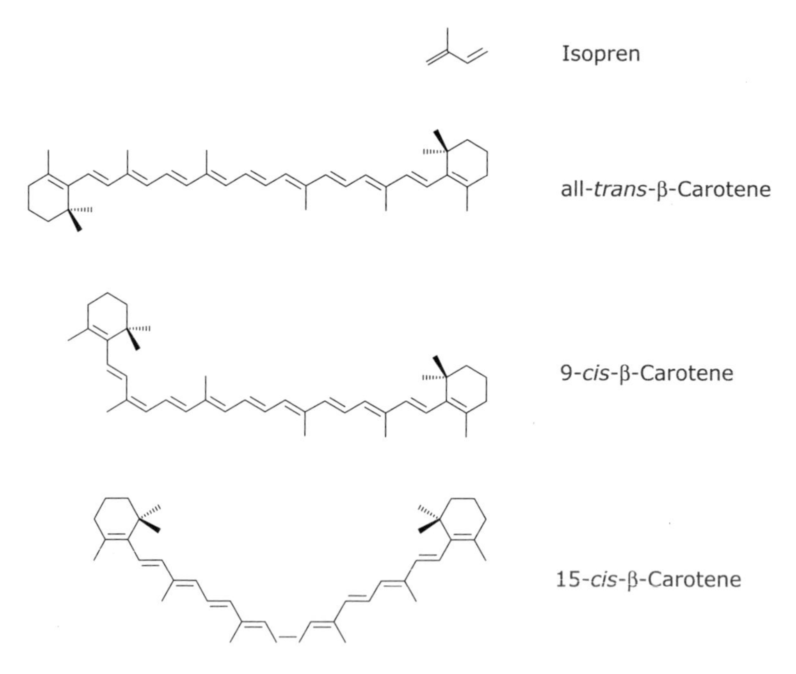

Fig. 2.8. Structures of β-carotenes. Structurally, these molecules are tetraterpenes, i.e. they are made of eight isoprene units.

2.2.4 Biological chromophores: nucleic acid

The absorption of UV light by nucleic acids arises from $n \rightarrow \pi^*$ and $\pi \rightarrow \pi^*$ transitions of the purine (adenine, guanine) and pyrimidine (cytosine, thymine, uracil) bases that occur between 260 and 275 nm (Table 2.3).

Table 2.3. Peak wavelengths and molar absorption coefficients of nucleic bases.

Chromophore	λ_{max} (nm)	ε (l mol^{-1} cm^{-1})
Adenine	261	13 400
Guanine	275	8 100
Cytosine	267	6 100
Thymine	265	7 900
Uracil	260	8 200

The absorption spectra of the bases in polymers (DNA) are sensitive to pH and are greatly influenced by electronic interactions between bases.

The UV absorbance of DNA increases with increasing temperature; the UV/Vis spectrum of DNA at elevated temperatures is very similar to the UV/Vis spectrum of the same DNA digested with restriction enzymes. This indicates that upon thermal denaturation, nucleotides are increasingly exposed to incident light, which is a similar situation as in mixtures of mononucleotides. Since less UV light is absorbed by DNA when it is in its native state, this phenomenon is known as the hypochromicity of DNA (see Section 2.2.11). The sequence of a DNA molecule affects its hypochromic effect. Due to the hypochromic effect, the UV absorbance of a DNA molecule changes upon thermal denaturation describing a sigmoid curve. The point of inflection is defined as the transition temperature (or melting temperature; T_m) of the sample. The T_m value is directly proportional to the content of guanine and cytosine bases in the DNA sample. As this parameter characterises the denaturing/renaturing of DNA strands, it has practical value for the design of oligonucleotides ('primers') in a polymerase chain reaction and other molecular biological experiments.

2.2.5 Organic chromophores: polyenes

Organic molecules with sufficient conjugated double bonds, i.e. extended π electron systems, constitute UV/Vis chromophores. The more extended the π electron system, the lower the HOMO-LUMO gap energy and thus the longer the peak wavelength λ_{max} and the larger the molar absorption coefficient (Table 2.4).

2.2.6 Inorganic chromophores

Inorganic molecules containing double bonds can absorb light in the UV/Vis region, typically due to $n \rightarrow \pi^*$ transitions (Table 2.5).

Table 2.4. Peak wavelengths and molar absorption coefficients of polyenes with an increasing number of double bonds.

Chromophore	N	λ_{max} (nm)	ε (l mol^{-1} cm^{-1})
	1	165	100
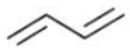	2	217	200
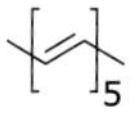 5	5	340	1 250
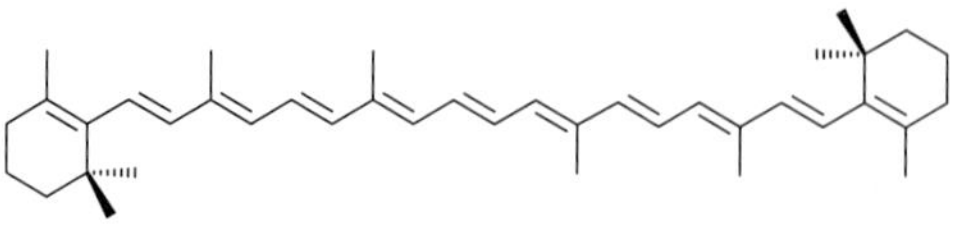	11	451	139 170

Table 2.5. Spectral characteristics of some inorganic anions in aqueous solution of their sodium salts in the wavelength range of 200–700 nm. All listed anions have very strong absorption bands below 200 nm.

Molecule	λ_{max} (nm)	ε (l mol^{-1} cm^{-1})
HCO_3^-	264	<1.0
NO_3^-	302	7.4
PO_4^{3-}	280	<1.0
SO_4^{2-}	267	<1.0
	222	<1.0
SCN^-	216	15 000

2.2.7 Chromophores by charge-transfer complexes

Many metal ions absorb light in the UV/Vis energetic range due to electronic transitions between *d* orbitals (*d-d* electronic transitions; this happens when the *d* orbitals split into energetically different sets as predicted by the Ligand Field Theory). Light of the wavelength corresponding to the energy difference between two *d* orbitals is absorbed, thus promoting an electron from a lower *d* orbital into a higher orbital of the *d* shell. However, transitions between *d* orbitals are quantum mechanically forbidden (including the rule $\Delta S = 0$); therefore these transitions give rise to colouring of low intensity. Remember that intensity is the macroscopic manifestation of quantum mechanical probabilities.

In complexes with ligands, metal ions can form charge-transfer complexes. For a complex to demonstrate charge-transfer behaviour, one of its components must have electron-donating properties and another component must be able to accept electrons. Absorption of radiation initiates the transfer of an electron from the donor to an orbital associated with the acceptor. As these transitions do not violate quantum mechanical rules, they give rise to intense colouring; the molar absorption coefficients from charge-transfer absorption are thus large, typically $\varepsilon > 10000\,l\,mol^{-1}\,cm^{-1}$. The redistribution of electrons results in differences in the solvation for the ground-state and excited-state molecules. Accordingly, the energy of charge-transfer bands is highly dependent on solvent.

Charge-transfer complexes are not limited to metal-ion complexes; they can occur between any two molecules forming a complex if appropriate electron donor and acceptor properties are present. A prominent example is the complex formed between iodine and starch, which yields an intense purple colour. As bank notes are not sized with starch, the application of iodine solution to a specimen can serve as a rough screen for counterfeit currency.

Ligand-to-metal charge-transfer complexes

Ligand-to-metal charge-transfer (LMCT) complexes arise from transfer of electrons from molecular orbitals (see Section 1.1.2) with ligand-like character to those with metal-like character. Predominantly, this type of charge transfer is observed with ligands that possess lone electron pairs with relatively high energy (such as sulfur or selenium) or if the metal has low-lying empty orbitals. In many such complexes, the metal is present in a high oxidation state (Table 2.6).

Table 2.6. Examples of LMCT complexes. The listed pigments appear coloured because they only reflect light of the visible colour. Light with a wavelength of the complementary colour is absorbed due to ligand-to-metal charge transfer.

Complex	λ_{max} (nm)	Absorbed colour	Visible colour	Electron transfer
MnO_4^-	565	Yellow	Purple	O^{2-} $(2p) \rightarrow Mn^{7+}$ $(2d[x^2 - y^2], 2d[z^2])$[a]
	340	Near-UV	Invisible	O^{2-} $(2p) \rightarrow Mn^{7+}$ $(2d[xy], 2d[xz], 2d[yz])$[a]
CdS	480	Blue	Yellow	S^{2-} $(3p) \rightarrow Cd^{2+}$ $(5s)$
HgS	365	Violet	Red	S^{2-} $(3p) \rightarrow Hg^{2+}$ $(6s)$
Fe oxides	~400	Blue-violet	Red and yellow	O^{2-} $(2p) \rightarrow Fe^{n+}$ $(3d)$

[a] In the presence of ligands, the five degenerated *d* orbitals split into energetically higher- and lower-lying orbitals according to the coordination geometry (Ligand Field Theory).

Table 2.7. Examples of metal-to-ligand charge-transfer complexes and their applications. Note that the compounds below have rather broad absorption bands with local maxima at the given λ_{max}; therefore, the visible colour is not the one expected from the given λ_{max} value.

Complex	λ_{max} (nm)	Visible colour	Application
$[Ru(bipy)_3]^{2+}$	452	Orange	Versatile photochemical RedOx reagent with a lifetime of microseconds
$W(CO)_4(phen)$	546	Red	Photochemical reactive reagent
$[Fe(phen)_3]^{2+}$	508	Orange	Colorimetric analysis of iron
$[Fe(bipy)_3]^{2+}$	508	Orange	Colorimetric analysis of iron

Carbon monoxide Cyanide Thiocyanate (rhodanide) 2,2-Bipyridine 1,10-Phenanthroline

Fig. 2.9. Examples of common ligands in metal-to-ligand charge-transfer complexes.

Metal-to-ligand charge-transfer complexes

In metal-to-ligand charge-transfer complexes, the electrons transition from metal-like orbitals to empty ligand-like orbitals. Such transitions are most commonly observed with ligands that possess low-lying π^* orbitals. Therefore, these ligands typically possess conjugated double or triple bonds, such as aromatic ligands (Fig. 2.9). The electronic transition will occur at low energy if the metal ion has a low oxidation number.

Examples of these complexes are shown in Table 2.7.

2.2.8 Quantification of light absorption

The chance for a photon to be absorbed by matter is given by the molar extinction coefficient ε, which itself is dependent on the wavelength λ of the photon.

If light with the intensity I_0 passes a sample with appropriate transparency and the thickness d, the intensity I drops along the pathway in an exponential manner. The characteristic absorption parameter for the sample is α; this yields the correlation:

$$I = I_0 e^{-\alpha d}. \quad (2.1)$$

The quotient

$$T = \frac{I}{I_0} \tag{2.2}$$

is called the transmittance, thus yielding

$$lnT = -\alpha d. \tag{2.3}$$

Biochemical samples usually comprise aqueous solutions, where the substance of interest is present at a molar concentration c. If one transforms the exponential (logarithmic) correlation into an expression based on the decadic logarithm, the resulting expression is the rule of Lambert–Beer, with ε being the molar absorption coefficient (molar extinction coefficient):

$$I = I_0 \times 10^{-\varepsilon cd} \tag{2.4}$$

where $[d] = 1$ cm, $[c] = 1$ mol l^{-1} and $[\varepsilon] = 1\,l$ mol^{-1} cm^{-1}.

This yields:

$$\log\frac{I_0}{I} = \log\frac{1}{T} = \varepsilon \times c \times d = A \tag{2.5}$$

where A is the absorbance of the sample, which is displayed by the spectrophotometer.

The rule of Lambert–Beer is valid for low concentrations only. Higher concentrations might lead to association of molecules and therefore cause deviations from the Lambert–Beer behaviour. In a sample with more than one absorbing species, the absorbance is an additive parameter (as well as the extinction coefficient).

Note that scattering effects (especially if the sample contains a dispersion or suspension) increase the absorbance artificially, as the scattered light is missing from the transmitted light when it hits the detector.

2.2.9 Absorption or light scattering: optical density

In some applications, for example measurement of turbidity of cell cultures (determination of biomass concentration), it is not the absorption but the scattering of light (see Section 2.6) that is actually measured with a spectrophotometer. Extremely turbid samples like bacterial cultures do not absorb the incoming light. Instead, the light is scattered and thus the spectrometer will record an apparent absorbance (sometimes also called attenuance). In this case, the observed parameter is called optical density (OD). Instruments specifically designed to measure turbid samples are nephelometers or Klett meters; however, most biochemical laboratories use the general UV/Vis spectrometer for determination of optical densities of cell cultures.

2.2.10 Deviations from the Lambert–Beer rule

According to Lambert–Beer, absorbance is linearly proportional to the concentration of chromophores. This may no longer be the case in samples with high absorbances. Every spectrophotometer has a certain amount of stray light, which is light received at the detector but not anticipated in the spectral band isolated by the monochromator. In order to obtain reasonable signal-to-noise ratios, the intensity of light at the chosen wavelength (I_λ) should be at least ten times higher than the intensity of the stray light (I_S). If the stray light gains in intensity, the effects measured in the detector have nothing or little to do with chromophore concentration anymore. Secondly, molecular events might lead to deviations from the Lambert–Beer rule. For instance, chromophores might dimerise at high concentrations and the resulting chromophore might possess different spectroscopic parameters.

2.2.11 Factors affecting UV/Vis absorption

Biochemical samples usually are in buffered aqueous solution, which has two major advantages:

- Proteins and peptides are stable in water as a solvent, which is also the 'native' solvent.
- In the wavelength interval of UV/Vis (700–200 nm) the water spectrum does not show any absorption bands and thus acts as a silent component in the sample.

The absorption spectrum of a chromophore is only partly determined by its chemical structure. The environment affects the observed spectrum, which mainly can be described by three parameters:

- protonation/deprotonation (pH, RedOx);
- solvent polarity (dielectric constant of the solvent); and
- orientation effects.

Vice versa, the immediate environment of chromophores can be probed by assessing their absorbance, which makes chromophores ideal reporter molecules for environmental factors.

Four effects, two each for wavelength and absorption changes, have to be considered. A wavelength shift to higher values is called a red shift or bathochromic effect; similarly, a shift to lower wavelengths is called a blue shift or hypsochromic effect. An increase in absorption is called

Table 2.8. Dielectric constants of vacuum and some solvents at 25 °C.

Medium	Dielectric constant ε
Vacuum	1.00
Dioxane	2.20
CH_2Cl_2	8.51
Ethanol	24.3
Glycerol	42.5
DMSO	47.2
H_2O	78.4

hyperchromicity ('more colour'), while a decrease in absorption is called hypochromicity ('less colour').

Protonation/deprotonation arises either from changes in pH or from oxidation/reduction reactions, which makes chromophores pH- and RedOx-sensitive reporters. For instance, with amino acids, an increase in λ_{max} (bathochromic effect) as well as absorbance (hyperchromicity) is observed when a titratable group such as tyrosine becomes charged.

Similarly, solvents of different polarities lead to different absorption spectra. Solvent polarity is measured by the dielectric constant ε (see Table 2.8), and affects the difference between the ground and excited states of molecules – an effect that is due to different stabilisation of the excited state. As a general rule, when shifting to a less polar environment one observes a bathochromic effect. Conversely, a solvent with higher polarity elicits a hypsochromic effect. The absorption maximum of the free amino acid tryptophan, for example, shows a bathochromic shift (red shift) from 279.8 nm in water to 281.3 nm in ethanol, a solvent of less polarity. Similarly, caffeine dissolved in either water or methylene chloride exhibits a bathochromic shift when changing to the less polar solvent (Fig. 2.10).

Lastly, orientation effects, such as an increase in order of nucleic acids from single-stranded to double-stranded DNA, lead to different absorption behaviour. A sample of free nucleotides exhibits a higher absorption than a sample with identical amounts of nucleotides but assembled into a single-stranded polynucleotide. Accordingly, double-stranded polynucleotides exhibit an even smaller absorption than two single-stranded polynucleotides. This phenomenon is called the hypochromicity of polynucleotides. The increased exposure (and thus stronger absorbance) of the individual nucleotides in the less ordered states provides a simplified explanation for this behaviour.

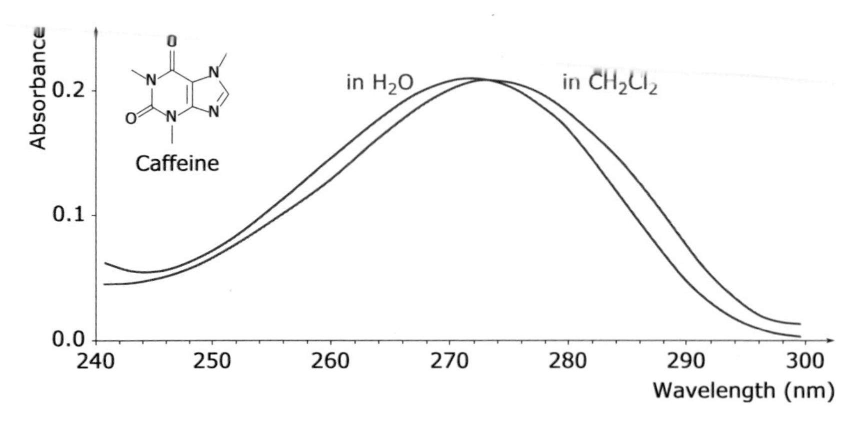

Fig. 2.10. UV/Vis spectra of the xanthine caffeine in water and methylene chloride. An aqueous solution of caffeine displays a λ_{max} of 272 nm. In methylene chloride, a solvent of less polarity than water, λ_{max} shifts to 274 nm – a bathochromic effect.

2.2.12 UV/Vis measurements

UV/Vis spectrophotometers are usually dual-beam spectrometers where the first channel contains the sample and the second channel holds the control (buffer) for subtraction. Alternatively, one can record the control spectrum first and use this as an internal reference for the sample spectrum. The latter approach has become very popular, as many spectrometers in laboratories are computer controlled, and baseline correction can be carried out using the software by simply subtracting the control from the sample spectrum, eliminating the need for matched sample cells.

The light source is a tungsten filament bulb for the visual part of the spectrum and a deuterium bulb for the UV region. As the emitted light consists of many different wavelengths, a monochromator, consisting of either a prism or a rotating metal grid of high precision called a grating, is placed between the light source and the sample. Wavelength selection can also be achieved by using coloured filters that absorb all but a certain limited range of wavelengths. This limited range is called the bandwidth of the filter. Filter-based wavelength selection is used in colorimetry, a method with moderate accuracy, but is best suited for specific colorimetric assays where only certain wavelengths are of interest. If wavelengths are selected by prisms or gratings, the technique is called spectrophotometry (Fig. 2.11).

A prism splits the incoming light into its components by refraction. Refraction occurs because radiation of different wavelengths travels along different paths in medium of higher density. In order to maintain the principle of velocity conservation, light of shorter wavelength (higher

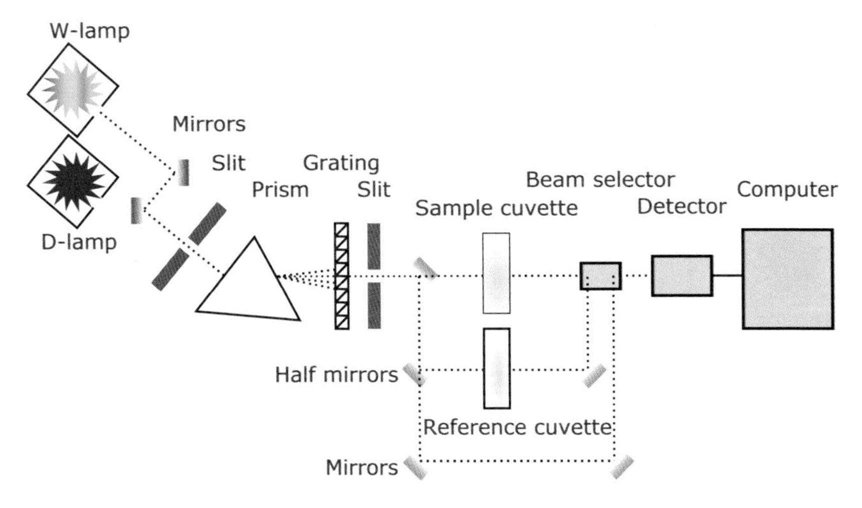

Fig. 2.11. Optical arrangements in a dual-beam spectrophotometer. Either a prism or a grating constitutes the monochromator of the instrument. Optical paths are shown as dotted lines.

frequency) must travel a longer distance (cf. the blue sky effect). At a grating, the splitting of wavelengths is achieved by diffraction. Diffraction is a reflection phenomenon that occurs at a grid surface, in this case a series of engraved fine lines. The distance between the lines has to be of the same order of magnitude as the wavelength of the diffracted radiation. By varying the distance between the lines, different wavelengths are selected. This is achieved by rotating the grating perpendicular to the optical axis. The resolution achieved by gratings is much higher than the one available by prisms. Contemporary high-precision instruments almost exclusively contain gratings as monochromators, as they can be reproducibly made at high quality by photoreproduction.

Theoretically, a single wavelength is selected by the monochromator in spectrophotometers, and the emergent light is a parallel beam. Here, the bandwidth is defined as twice the half-intensity bandwidth. The bandwidth is a function of the optical slit width. The narrower the slit width, the more reproducible are the measured absorbance values. In contrast, the sensitivity becomes less as the slit narrows, because less radiation passes through to the detector.

In a dual-beam instrument, the incoming light beam is split into two parts by a half mirror. One beam passes through the sample and the other through a control (blank or reference). This approach obviates any problems of variation in light intensity, as both reference and sample would be affected equally. The measured absorbance is the difference between the two

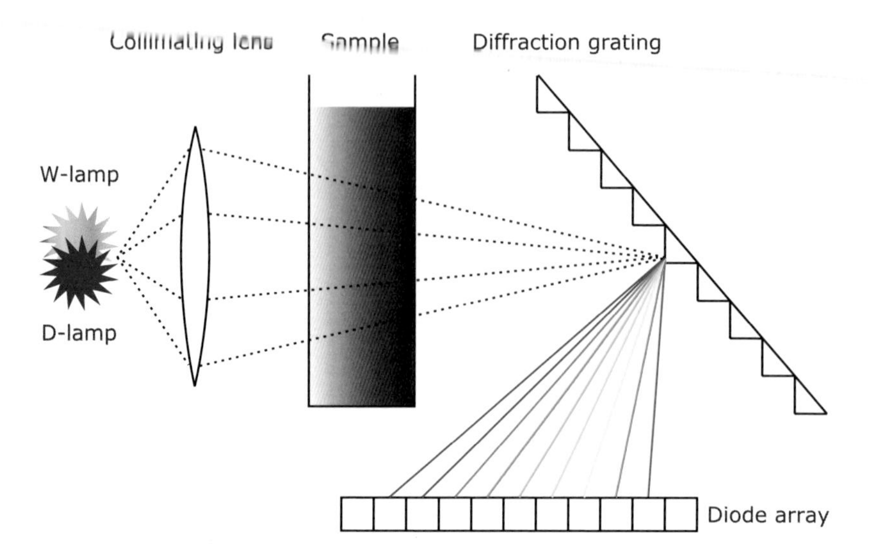

Fig. 2.12. Schematics of a UV/Vis spectrometer with diode array detector. Polychromatic light is used to illuminate the sample; individual wavelengths are separated after the light has passed through the sample.

transmitted beams of light recorded. Depending on the instrument, a second detector measures the intensity of the incoming beam, although some instruments use an arrangement where one detector measures the incoming and the transmitted intensity alternately. The latter design is better from an analytical point of view, as it eliminates potential variations between the two detectors. At about 350 nm, most instruments require a change of the light source from visible to UV light. This is achieved by mechanically moving mirrors that direct the appropriate beam along the optical axis and divert the other. When scanning the interval of 500–210 nm, this frequently gives rise to an offset of the spectrum at the switch-over point.

As borosilicate glass and normal plastics absorb UV light, such sample cells can only be used for applications in the visible range of the spectrum (to about 350 nm). For UV measurements, quartz cuvettes need to be used. However, disposable plastic cuvettes have been developed that allow measurements over the entire range of the UV/Vis spectrum.

In recent years, UV spectrometers with diode array detectors have also become popular (Fig. 2.12). The main difference from the conventional instruments described above is that here the sample is illuminated with polychromatic light, and the separation into individual wavelengths occurs after the light has passed through the sample. A diode array contains in the order of 1000 individual sensors in an array of 3–7 cm length. Typically, no slits are used and thus there is no intensity lost in the monochromator. To improve precision, multiple scans are acquired and averaged.

A colorimeter uses a coloured filter as the monochromator; the bandwidth of a colorimeter is thus determined by the filter. A filter that appears red to the human eye is transmitting red light and absorbs almost any other (visual) wavelength. This filter would be used to examine blue solutions, as these would absorb red light. The filter used for a specific colorimetric assay is thus made of a colour complementary to that of the solution being tested.

The usual procedure for (colorimetric) assays is to prepare a set of standards and produce a plot of concentration versus absorbance called a calibration curve. This should be linear as long as the Lambert–Beer law applies. Absorbances of unknowns are then measured and their concentration interpolated from the linear region of the plot. It is important that one never extrapolates beyond the region for which an instrument has been calibrated, as this potentially introduces enormous errors.

To obtain good spectra, the maximum absorbance should be approximately 0.1–0.5, which corresponds to concentrations of about 10–50 μM (assuming $\varepsilon = 10\,000\,\mathrm{l\,mol^{-1}\,cm^{-1}}$).

2.2.13 Qualitative and quantitative analysis

Qualitative analysis may be performed in the UV/Vis regions to identify certain classes of compounds both in the pure state and in biological mixtures (e.g. bound to protein). The application of UV/Vis spectroscopy to further analytical purposes is rather limited, but is possible for systems where appropriate features and parameters are known.

Most commonly, this type of spectroscopy is used for quantification of biological samples either directly or via colorimetric assays. In many cases, proteins can be quantified directly using their intrinsic chromophores, tyrosine and tryptophan. Protein spectra are acquired by scanning from 500 to 210 nm. The characteristic features in a protein spectrum are a band at 278/280 nm (Fig. 2.13) and at ~220 nm. The region from 500 to 300 nm provides valuable information about the presence of any prosthetic groups or coenzymes. Protein quantification by single wavelength measurements at 280 and 260 nm only should be avoided, as the presence of larger aggregates (contaminations or protein aggregates) gives rise to considerable Rayleigh scatter that needs to be corrected for (Fig. 2.13).

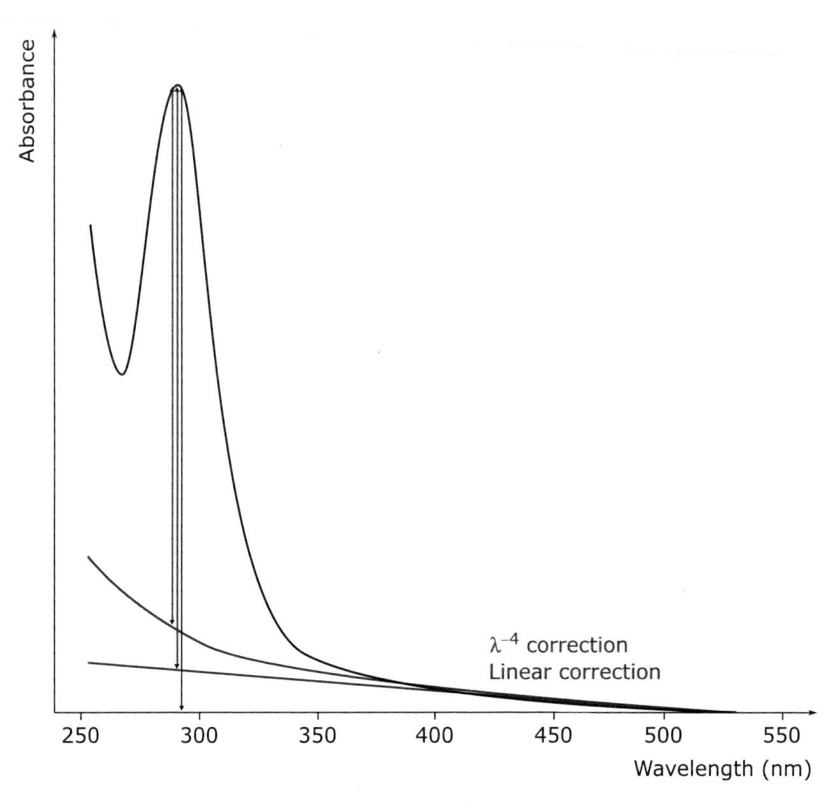

Fig. 2.13. The presence of larger aggregates in biological samples gives rise to Rayleigh scatter visible by a considerable slope in the region from 500 to 350 nm. The blue line shows the correction to be applied to spectra with Rayleigh scatter, which increases with λ^{-4}. Practically, in many cases, a linear extrapolation of the region from 500 to ~350 nm is performed to correct for the scatter (red).

2.2.14 Difference spectra

The main advantage of difference spectroscopy is its capacity to detect small absorbance changes in systems with high background absorbance. A difference spectrum is obtained by subtracting one absorption spectrum from another. Difference spectra can be obtained in two ways: either by subtraction of one absolute absorption spectrum from another, or by placing one sample in the reference cuvette and another in the test cuvette. The latter method requires use of a dual-beam instrument, while the former method has become very popular due to most instruments being controlled by computers, which allows easy processing and handling of data. From a purist's point of view, the direct measurement of the difference spectrum in a dual-beam instrument is the preferred method, as it reduces the introduction of inconsistencies between samples and thus the error of the measurement.

Difference spectra have three distinct features compared with absolute spectra:

- Difference spectra may contain negative absorbance values.
- Absorption maxima and minima may be displaced and the extinction coefficients are different from those in peaks of absolute spectra.
- There are points of zero absorbance, usually accompanied by a change of sign of the absorbance values. These points are observed at wavelengths where both species of related molecules exhibit identical absorbances (isosbestic points), and which may be used for checking for the presence of interfering substances.

Common applications for difference UV spectroscopy include the determination of the number of aromatic amino acids exposed to solvent, detection of conformational changes occurring in proteins, detection of aromatic amino acids in active sites of enzymes, and monitoring of reactions involving 'catalytic' chromophores (prosthetic groups, coenzymes).

2.2.15 Derivative spectroscopy

Another way to resolve small changes in absorption spectra that otherwise would remain invisible is the use of derivative spectroscopy. Here, the absolute absorption spectrum of a sample is differentiated and the differential $\delta^x A/\delta\lambda^x$ plotted against the wavelength. As the algebraic relationship between A and λ is unknown, differentiation is carried out by numerical methods using computer software. The usefulness of this approach depends on the individual problem. Examples of successful applications include the binding of a monoclonal antibody to its antigen with second-order derivatives and the quantification of tryptophan and tyrosine residues in proteins using fourth-order ($x = 4$) derivatives.

2.2.16 Solvent perturbation

As we have mentioned above, aromatic amino acids are the main chromophores of proteins in the UV region of the electromagnetic spectrum. Furthermore, the UV absorption of chromophores depends largely on the polarity in its immediate environment. A change in the polarity of the solvent can change the UV spectrum of a protein by bathochromic or hypsochromic effects without changing its conformation. This phenomenon is called solvent perturbation and can be used to probe the surface of a protein molecule. In order to be accessible to the solvent, the chromophore has to be accessible on the protein surface. Practically, solvents like dimethyl sulfoxide (DMSO), dioxane, glycerol, mannitol, sucrose and polyethylene glycol are used for solvent perturbation experiments, because they are miscible with water. The method of solvent perturbation is most commonly used for determination of the number of aromatic residues that are exposed to solvent.

2.2.17 Spectrophotometric and colorimetric assays

For biochemical assays testing for time- or concentration-dependent responses of systems, an appropriate read-out is required that is coupled to the progress of the reaction (reaction coordinate). Therefore, the biophysical parameter being monitored needs to be coupled to the biochemical parameter under investigation. Frequently, the monitored parameter is the absorbance of a system at a given wavelength, which is monitored throughout the course of the experiment. Preferably, one should try to monitor the changing species directly (e.g. protein absorption, starting product or generated product of a reaction), but in many cases this is not possible and a secondary reaction has to be used to generate an appropriate signal for monitoring. A common application of the latter approach is the determination of protein concentration by Lowry or Bradford assays, where a secondary reaction is used to colour the protein. The more intense the colour, the more protein is present. These assays are called colorimetric assays and a number of commonly used ones are listed in Table 2.9.

2.2.18 UV/Vis spectroscopic determination of protein concentration

The concentration of purified protein in aqueous solution can easily be assessed by recording a UV/Vis spectrum. The law of Lambert–Beer provides a relation between the spectroscopic absorbance and the concentration of a soluble substance causing this absorption:

$$A = \varepsilon \times c \times l \Rightarrow c = \frac{A}{\varepsilon \times l} \Rightarrow \rho^* = \frac{A \times M}{\varepsilon \times l}. \tag{2.6}$$

In the equations above, A stands for absorbance, c for the molar concentration, ε for the molar absorption (extinction) coefficient, l for the cuvette thickness, M for the molar mass and ρ^* for the mass concentration. Note that A and ε are wavelength dependent. Provided ε is known, the protein concentration can be calculated from the absorption.

Table 2.9. Common colorimetric and UV absorption assays. Luciferase assays are luminometric methods and discussed in Section 2.4.

Substance	Reagent	λ_{max} (nm)
Metal ions	Complexation → charge-transfer complex	
Amino acids	(a) Ninhydrin	570 (proline: 420)
	(b) Cupric salts	620
Cysteine residues, thiolates	Ellman reagent (di-sodium-bis-(3-carboxy-4-nitrophenyl)-disulfide)	412
Protein	(a) Folin (phosphomolybdate, phosphotungstate, cupric salt)	660
	(b) Biuret (reacts with peptide bonds)	540
	(c) BCA (bicinchoninic acid) reagent	562
	(d) Coomassie Brilliant Blue	595
	(e) Direct	Tyr, Trp: 278 Peptide bond: 190
Coenzymes	Direct	FAD: 438 NADH: 340 NAD^+: 260
Carotenoids	Direct	420, 450, 480
Porphyrins	Direct	400 (Soret band)
Carbohydrate	(a) Phenol, H_2SO_4	Glucose: 490 Xylose: 480
	(b) Anthrone (anthrone, H_2SO_4)	620 or 625
Reducing sugars	Dinitrosalicylate, alkaline tartrate buffer	540
Pentoses	(a) Bial (orcinol, ethanol, $FeCl_3$, HCl)	665
	(b) Cysteine, H_2SO_4	380–415
Hexoses	(a) Carbazole, ethanol, H_2SO_4	540 or 440
	(b) Cysteine, H_2SO_4	380–415
	(c) Arsenomolybdate	500–570
Glucose	Glucose oxidase, peroxidase, o-dianisidine, phosphate buffer	420
Ketohexose	(a) Resorcinol, thiourea, ethanoic acid, HCl	520
	(b) Carbazole, ethanol, cysteine, H_2SO_4	560
	(c) Diphenylamine, ethanol, ethanoic acid, HCl	635
Hexosamines	Ehrlich (dimethylaminobenzaldehyde, ethanol, HCl)	530
DNA	(a) Diphenylamine	595
	(b) Direct	260
RNA	Bial (orcinol, ethanol, $FeCl_3$, HCl)	665
Sterols and steroids	Liebermann–Burchardt reagent (acetic anhydride, H_2SO_4, chloroform)	425, 625
Cholesterol	Cholesterol oxidase, peroxidase, 4-amino-antipyrine, phenol	500
ATPase assay	Coupled enzyme assay with ATPase, pyruvate kinase, lactate dehydrogenase: ATP → ADP (consumes ATP) Phosphoenolpyruvate → pyruvate (consumes ADP) Pyruvate → lactate (consumes NADH)	NADH: 340

Table 2.10. Increment system for molar absorption coefficient according to Gill & von Hippel (Gill & von Hippel, 1989).

Residue	ε increment at 280 nm ($l\ mol^{-1}\ cm^{-1}$)
Cys_2	120
Trp	5690
Tyr	1280

Table 2.11. Increment system for molar absorption coefficient according to Mach *et al.* (1992).

Residue	ε increment ($l\ mol^{-1}\ cm^{-1}$)
Cys_2	134
Trp	5540
Tyr	1480

If the molar extinction coefficient has not been determined for a certain protein, it can be predicted by assuming an incremental ε value for each absorbing protein residue. Summation over all residues yields a reasonable estimation for the extinction coefficient. The simplest increment system is based on the values of Gill & von Hippel (1989).

This method only requires an absorption value at $\lambda = 280$ nm to calculate the protein concentration. The ε increments from Table 2.10 are used to calculate a molar extinction coefficient for the entire protein by summation over all relevant residues in the protein:

$$\varepsilon = \Sigma\, \varepsilon_{res} \times N_{res}. \tag{2.7}$$

This calculation is the most popular approach and is implemented in many software programs such as, for example, EMBOSS Pepstats (http://www.ebi.ac.uk/Tools/seqstats/emboss_pepstats/) (Rice *et al.*, 2000) or Peptides (Hofmann & Wlodawer, 2002). The concentration can then be calculated using the formula derived from the law by Lambert–Beer:

$$\rho^* = \frac{A \times M}{\varepsilon \times l}. \tag{2.8}$$

A more complicated increment system was introduced by Mach *et al.* (1992), which requires absorption values at the wavelengths 280, 320 and 350 nm.

The ε increments from Table 2.11 are used to calculate a molar extinction coefficient for the entire protein by summation over all relevant residues in the protein. The mass concentration of the protein is then calculated according to:

$$\rho^* = \frac{(A_{280} - 10^{2.5 \times \log A_{320}} - 1.5 \times A_{350}) \times M}{\varepsilon \times l}. \tag{2.9}$$

If the molar extinction coefficient is not known or cannot be estimated, the empirical formula of Warburg-Christian (Warburg & Christian, 1941) provides a simple way of determining protein concentrations:

$$\rho^* = (1.55 \times A_{280} - 0.76 \times A_{260}) \text{ mg ml}^{-1} \tag{2.10}$$

Here, the absorption at $\lambda = 280$ nm and $\lambda = 260$ nm has to be determined in order to calculate the concentration.

2.2.19 UV/Vis spectroscopic determination of DNA concentration

DNA concentration can be determined spectroscopically (Sambrook *et al.*, 1989). Assuming an averaged molecular mass of $M = 500$ g mol^{-1} for the nucleotides, the absorbance A of a DNA solution at 260 nm can be converted to mass concentration ρ^* by:

$$\rho^* = 50 \text{ µg ml}^{-1} \times A_{260} = 0.05 \text{ µg µl}^{-1} \times A_{260} \tag{2.11}$$

The ratio A_{260}/A_{280} is an indicator for the purity of the DNA solution and should range between 1.8 and 2.0. However, due to the much higher molar absorption coefficients at 260 and 280 nm of nucleic acids compared with proteins, even rather high protein concentrations affect this ratio only marginally. For the same reason, DNA contamination of a protein sample is readily seen from the ratio A_{280}/A_{260}, a pure protein sample is supposed to possess a ratio between 1.8 and 2.0.

2.3 Fluorescence spectroscopy

2.3.1 General remarks

Fluorescence is an emission phenomenon where an energy transition from a higher to a lower state is accompanied by radiation. Only molecules in their excited forms are able to emit fluorescence; thus, they have to be brought into a state of higher energy prior to the emission phenomenon.

We have already seen in Section 1.1.2 that molecules possess discrete states of energy. The potential energy levels of molecules have been depicted by different Lennard-Jones potential curves with overlaid vibrational (and rotational) states (see Fig. 1.3). Such diagrams can be abstracted further to yield Jablonski diagrams (Fig. 2.14).

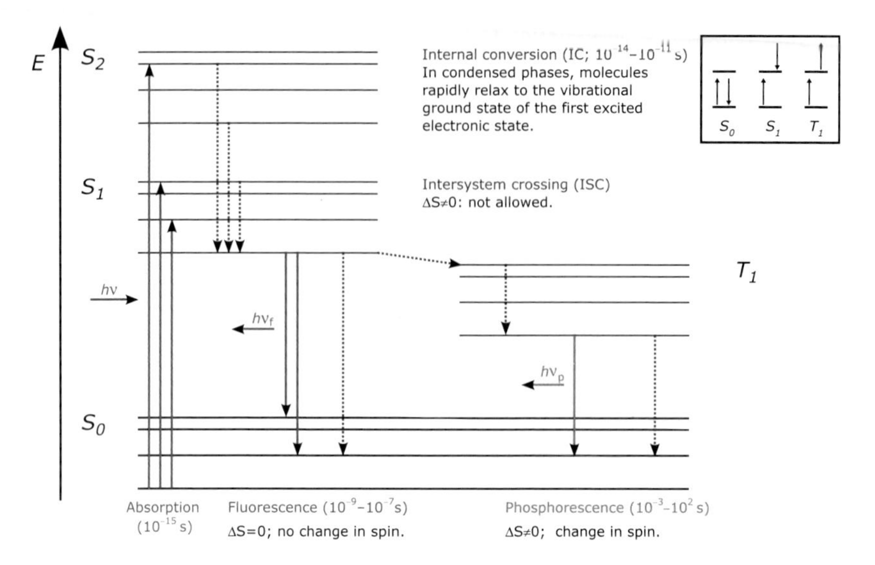

Fig. 2.14. Jablonski diagram. Shown are the electronic ground state (S_0), two excited singlet states (S_1, S_2) and a triplet state (T_1). Vibrational levels are only illustrated as exemplary. Solid vertical lines indicate radiative transitions, and dotted lines show non-radiative transitions. The inset shows the relationship between electron configurations, multiplicity M and total spin number S (see Equations 1.4 and 1.5, respectively). Internal conversion (IC, blue dotted transitions): in condensed phases, molecules rapidly relax to the vibrational ground state of the first excited electronic state. Intersystem crossing (ISC; red dotted transition): change from a singlet to a triplet state. This is a disallowed transition, as $\Delta S \neq 0$.

In these diagrams, energy transitions are indicated by vertical lines. Not all transitions are possible; allowed transitions are defined by the selection rules of quantum mechanics. A molecule in its electronic and vibrational ground state ($S_0 v_0$) can absorb photons matching the energy difference of its various discrete states. The required photon energy has to be higher than that required to reach the vibrational ground state of the first electronic excited state ($S_1 v_0$). The excess energy is absorbed as vibrational energy ($v > 0$), and quickly dissipated as heat by collision with solvent molecules. The molecule thus returns to the vibrational ground state ($S_1 v_0$). These relaxation processes are non-radiating transitions from one energetic state to another with lower energy, and are called internal conversion (IC). From the lowest level of the first electronic excited state, the molecule returns to the ground state (S_0) either by emitting light (fluorescence) or by a non-radiative transition. Upon radiative transition, the molecule can end up in any of the vibrational states of the electronic ground state (as per quantum mechanical rules; see also Table 2.14).

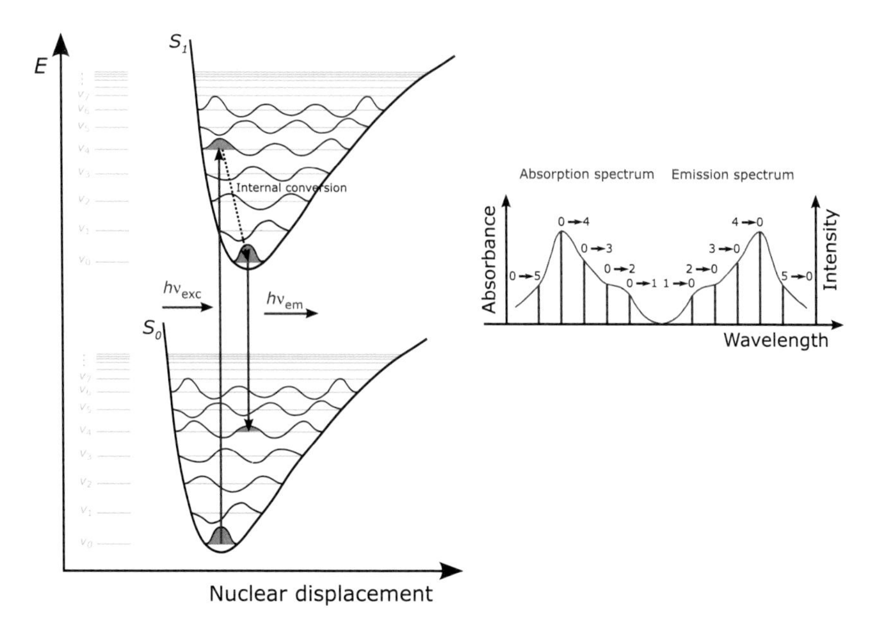

Fig. 2.15. The Franck–Condon principle determines the molecular transitions upon absorption and emission of light. Blue colour indicates absorption and red colour emission. Left: The wavelength λ_{max} for absorption of light is determined by the highest overlap integral (grey shaded areas) of wave functions for the vibrational states (green) in the electronic ground (S_0) and excited (S_1) states. In this example, the most likely transition occurs between $S_0 v_0$ and $S_1 v_2$. If the molecule is capable of fluorescence, the most likely transition upon emission is between $S_1 v_0$ and $S_0 v_2$, thus employing the same vibrational modes but in a different order. Right: The absorption/excitation and emission band spectra of molecules reflect the transition probabilities.

If the upper vibrational levels of the ground state overlap with those of the electronic excited state, the molecule will not emit fluorescence, but will rather revert to the ground state by non-radiative internal conversion. This is the most common way for excitation energy to be dissipated and is why fluorescent molecules are rather rare. Most molecules are flexible and thus have very high vibrational levels in the ground state. Indeed, most fluorescent molecules possess fairly rigid aromatic rings or ring systems. The fluorescent group in a molecule is called a fluorophore.

Comparison of the absorption (excitation) and fluorescence emission spectra shows that the fluorescence spectrum is shifted to lower wavelengths (Fig. 2.15). The light observed as fluorescence thus possesses lower energy than the light required to excite the molecule (Stokes shift). The reason for this phenomenon is that excitation of the molecule follows the Franck–Condon principle: firstly, the transition between two electronic states is dependent on the probability density given by the quantum mechanical wave functions (different for each vibration state). Secondly, the electronic transition is much quicker than the change of the atomic distance in the molecule. The atomic distance is thus considered constant and the transition line in Jablonski diagrams is vertical.

When the molecule absorbs a photon and transitions from the vibrational ground state in the electronic ground state (S_0v_0) to the excited state (S_1), a particular vibrational mode in the excited state will be the most populated state (e.g. S_1v_2; see Fig. 2.15). The selection of this mode is determined by the Franck–Condon factor, which describes the overlap integral of the quantum mechanical wave functions for each vibrational state. The larger the overlap between the wave functions of the vibrational modes of the electronic ground and excited states, the more likely a transition between the two will occur. Notably, the Franck–Condon factor is the same for absorption and fluorescence. In the chosen example, where photon absorption leads to a transition $S_0v_0 \rightarrow S_1v_2$, the most frequent fluorescence event will be $S_1v_0 \rightarrow S_0v_2$.

Through interactions with neighbouring molecules, the excited molecule may lose vibrational energy, but it will remain in the electronic excited state (S_1). The transition from the electronic excited state to the ground state will thus occur from the vibrational ground state of the electronic excited state (S_1v_0), again following the Franck–Condon principle. Similar to absorption bands, the intensity of the emitted light is proportional to the transition probability from the electronic excited to the ground state. Therefore, the band features of the absorption spectrum are due to vibrational levels in the electronic excited state, and the fluorescence spectrum reflects the vibrational levels of the electronic ground state.

The fluorescence properties of a molecule are determined by properties of the molecule itself (internal factors), as well as the environment of the molecule (external factors). The fluorescence intensity emitted by a molecule is dependent on the lifetime of the excited state (for details, see Section 2.3.5). The effective lifetime τ of excited molecules, however, differs from the fluorescence lifetime τ_r, as other non-radiative processes also affect the number of molecules in the excited state. τ is dependent on all processes that cause relaxation: fluorescence emission, internal conversion (IC), quenching (Q), fluorescence resonance energy transfer (FRET), reactions of the excited state and intersystem crossing (ISC).

The ratio of photons emitted and photons absorbed by a fluorophore is called the quantum yield Φ. It equals the ratio of the rate constant for fluorescence emission k_r to the sum of the rate constants for all six processes mentioned above:

$$\Phi = \frac{N_{\text{emitted}}}{N_{\text{absorbed}}} = \frac{k_r}{k} = \frac{k_r}{k_r + k_{IC} + k_{ISC} + k_{\text{reaction}} + k_Q c(Q) + k_{FRET}} = \frac{\tau}{\tau_r} \tag{2.12}$$

The quantum yield is a unitless quantity, and, most importantly, the only absolute measure of fluorescence of a molecule. Measuring the quantum yield is an elaborate process and requires comparison with a fluorophore of known quantum yield. In biochemical applications, this measurement is rarely done. Most commonly, the fluorescence emissions of two or more related samples are compared and their relative differences analysed.

An associated phenomenon in this context is phosphorescence, which arises from a transition from a triplet state (T_1) to the electronic (singlet) ground state (S_0). The molecule gets into the triplet state from an electronic excited singlet state by a process called intersystem crossing. The forbidden transition from singlet to triplet only happens with low probability in certain molecules where the electronic structure is favourable. Such molecules usually contain 'heavier' atoms, such as for example zinc or strontium, where the S_1 and T_1 states overlap energetically. The rate constants for phosphorescence are much longer and phosphorescence thus happens with a long delay and persists even when the exciting energy is no longer applied.

2.3.2 Fluorescence measurements

Fluorescence spectroscopy works most accurately at very low concentrations of emitting fluorophores. UV/Vis spectroscopy, in contrast, is least accurate at such low concentrations. One major factor adding to the high sensitivity of fluorescence applications is the spectral selectivity. Due to the Stokes shift, the wavelength of the emitted light is different from that of the exciting light. Another feature makes use of the fact that fluorescence is emitted in all directions. By placing the detector perpendicular to the excitation pathway, the background of the incident beam is reduced.

The schematics of a typical spectrofluorimeter are shown in Fig. 2.16. Two monochromators are used, one for tuning the wavelength of the exciting beam and a second for analysis of the fluorescence emission. Due to the emitted light always having a lower energy than the exciting light, the wavelength of the excitation monochromator is set at a lower

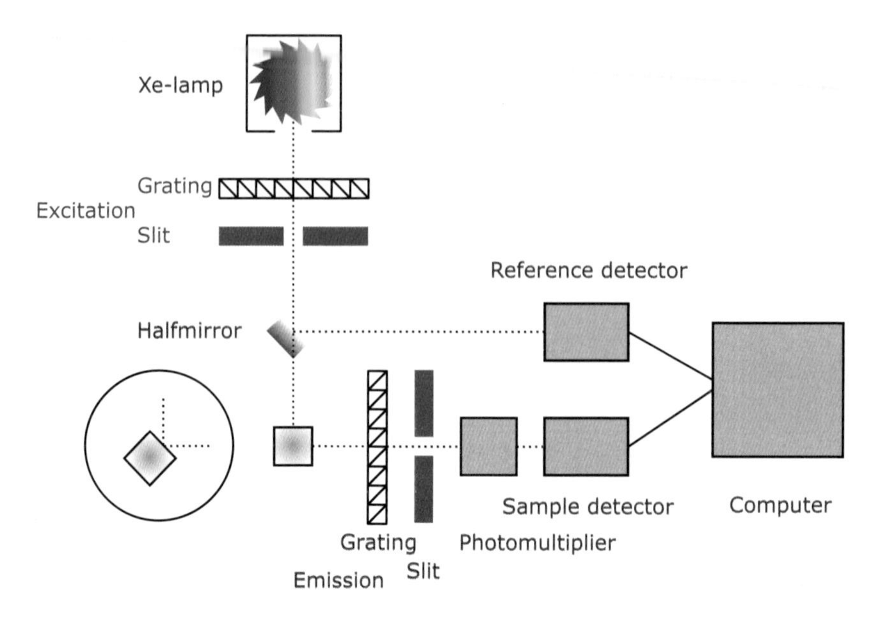

Fig. 2.16. Schematics of a spectrofluorimeter. Optical paths are shown as dotted lines. Inset: The geometry of front-face illumination.

wavelength than the emission monochromator. High-quality fluorescence spectrometers have a photon-counting detector yielding very high sensitivity. Temperature control is required for accurate work, as the emission intensity of a fluorophore is dependent on the temperature of the solution.

Two geometries are possible for the measurement, with the 90° arrangement most commonly used. Pre- and post-filter effects can arise owing to absorption of light prior to reaching the fluorophore and the reduction of emitted radiation. These phenomena are also called inner filter effects and are more evident in solutions with high concentrations. As a rough guide, the absorbance of a solution to be used for fluorescence experiments should be less than 0.05. The use of microcuvettes containing less material can also be useful. Alternatively, the front-face illumination geometry (Fig. 2.16, inset) can be used, which obviates the inner filter effect. Also, while the 90° geometry requires cuvettes with two neighbouring faces being clear (usually, fluorescence cuvettes have four clear faces), the front-face illumination technique requires only one clear face, as excitation and emission occur at the same face. However, front-face illumination is less sensitive than the 90° illumination.

Fluorescence spectra are always plots of fluorescence emission intensities against the wavelength (see for example Fig. 2.17). If the emission intensity at a single wavelength (detector) is scanned with variable excitation wavelength, the resulting spectrum is called an excitation spectrum.

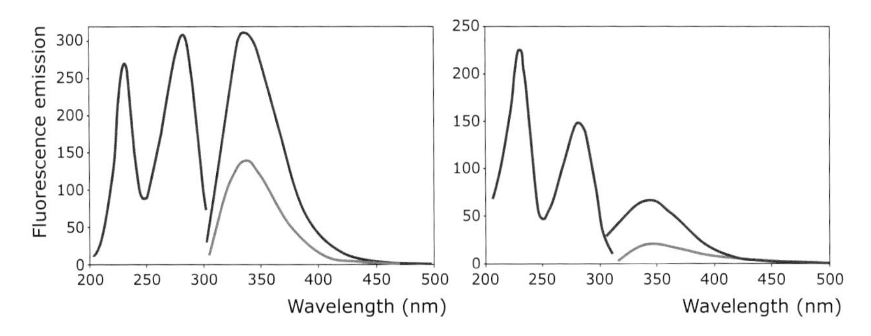

Fig. 2.17. Comparison of fluorescence excitation and emission spectra can yield insights into internal quenching. Shown are the fluorescence excitation spectra ($\lambda_{em} = 340$ nm; blue), and the emission spectra with $\lambda_{exc} = 280$ nm (red) and $\lambda_{exc} = 295$ nm (orange). Left: Phosphodiesterase homologue (*Escherichia coli*). Right: Cyclic phosphodiesterase (*Arabidopsis thaliana*). In the protein on the right, the fluorophores are located in close proximity to each other, which leads to the effect of intrinsic quenching, as obvious from the lower intensity of the emission spectrum compared with the excitation spectrum.

In contrast, if the excitation wavelength is set at a fixed value and the emission intensity is recorded for variable detection wavelengths, the resulting spectrum is called an emission spectrum. For a constant excitation energy, the excitation spectrum resembles the absorption spectrum (UV).

The UV absorption spectrum and the fluorescence emission spectrum of the same sample are more or less symmetrical with each other, with the fluorescence spectrum on the low-energy side ('red', longer wavelength) and the UV spectrum on the high-energy side ('blue', shorter wavelength).

2.3.3 Applications

There are many and highly varied applications for fluorescence, despite the fact that relatively few compounds exhibit the phenomenon. The effects of pH, solvent composition and the polarisation of fluorescence may all contribute to structural elucidation.

Measurement of fluorescence lifetimes can be used to assess rotation correlation coefficients and thus particle sizes. The fluorescence lifetime of a fluorophore can be assessed by time-correlated single-photon counting whereby the number of fluorescence (single-photon) events is counted at particular time intervals after an excitation flash. The resulting probability distribution can be fitted to an exponential function and the fluorescence lifetime deduced from its power.

Non-fluorescent compounds are often labelled with fluorescent probes to enable monitoring of molecular events. This is termed extrinsic fluorescence, as distinct from intrinsic fluorescence where the native compound exhibits the property. Some fluorescent dyes are sensitive to the presence of metal ions and can thus be used to track changes of these ions in samples *in vitro*, as well as in whole cells.

As fluorescence spectrometers have two monochromators, one for tuning the excitation wavelength and one for analysing the emission wavelength of the fluorophore, one can measure two types of spectra: excitation and emission spectra. For fluorescence excitation spectrum measurement, one sets the emission monochromator at a fixed wavelength (λ_{em}) and scans a range of excitation wavelengths which are then recorded as ordinate (x coordinate) of the excitation spectrum; the fluorescence emission at λ_{em} is plotted as the abscissa. Measurement of emission spectra is achieved by setting a fixed excitation wavelength (λ_{exc}) and scanning a wavelength range with the emission monochromator. To yield a spectrum, the emission wavelength λ_{em} is recorded as the ordinate and the emission intensity at λ_{em} is plotted as the abscissa.

Intrinsic protein fluorescence

Proteins possess three intrinsic fluorophores: tryptophan, tyrosine and phenylalanine, although the last has a very low quantum yield and its contribution to protein fluorescence emission is thus negligible (Table 2.12). Of the remaining two residues, tyrosine has the lower quantum yield and its fluorescence emission is almost entirely quenched when it

Table 2.12. Fluorescence properties of biological fluorophores in aqueous solution at pH 7.0.

Fluorophore	Absorption		Fluorescence emission			
	λ_{max} (nm)	ε (λ_{max}) (l mol^{-1} cm^{-1})	λ_{max} (nm)	Φ_F [a]	τ_0 [b] (ns)	ε (λ_{max}) $\times$ Φ_F [c] (l mol^{-1} cm^{-1})
Tryptophan	280	5 690	348	0.20	2.6	1138
Tyrosine	274	1 280	303	0.14	3.6	179
Phenylalanine	257	210	282	0.04	6.4	8
Adenine	260	13 400	321	2.6×10^{-4}	<0.02	3
Guanine	275	8 100	329	3.0×10^{-4}	<0.02	2
Cytosine	267	6 100	313	0.8×10^{-4}	<0.02	0.5
Uracil	260	9 500	308	0.4×10^{-4}	<0.02	0.4
NADH	340	6 200	470	0.02	0.40	118

[a] Quantum yield.

[b] Lifetime.

[c] The product of the absorption coefficient and quantum yield is called sensitivity.

becomes ionised, or is located near an amino or carboxyl group, or a tryptophan residue. Intrinsic protein fluorescence is thus usually determined by tryptophan fluorescence, which can be selectively excited at 295–305 nm. Excitation at 280 nm excites tyrosine and tryptophan fluorescence and the resulting spectra may therefore contain contributions from both types of residues.

The main application for intrinsic protein fluorescence aims at conformational monitoring. We have already mentioned that the fluorescence properties of a fluorophore depend significantly on environmental factors, including solvent, pH, possible quenchers (see Section 2.3.4) and neighbouring groups.

A number of empirical rules can be applied to interpret protein fluorescence spectra:

- As a fluorophore moves into an environment with less polarity, its emission spectrum exhibits a hypsochromic shift (λ_{max} moves to shorter wavelengths) and the intensity at λ_{max} increases.
- Fluorophores in a polar environment show a decrease in quantum yield with increasing temperature. In a non-polar environment, there is little change.
- Tryptophan fluorescence is quenched by neighbouring protonated acidic groups.

When interpreting effects observed in fluorescence experiments, one has to consider carefully all possible molecular events. For example, a compound added to a protein solution can cause quenching of tryptophan fluorescence. This could come about by binding of the compound at a site close to the tryptophan (i.e. the residue is surface exposed to a certain degree), or due to a conformational change induced by the compound.

The comparison of protein fluorescence excitation and emission spectra can yield insights into the location of fluorophores. The close spatial arrangement of fluorophores within a protein can lead to quenching of fluorescence emission; this might be seen by the lower intensity of the emission spectrum when compared with the excitation spectrum (Fig. 2.17).

Extrinsic fluorescence

Frequently, molecules of interest for biochemical studies are non-fluorescent. In many of these cases, an external fluorophore can be introduced into the system by chemical coupling or non-covalent binding. Some examples of commonly used external fluorophores are shown in Fig. 2.18. Three criteria must be met by fluorophores in this context. Firstly, the molecule must

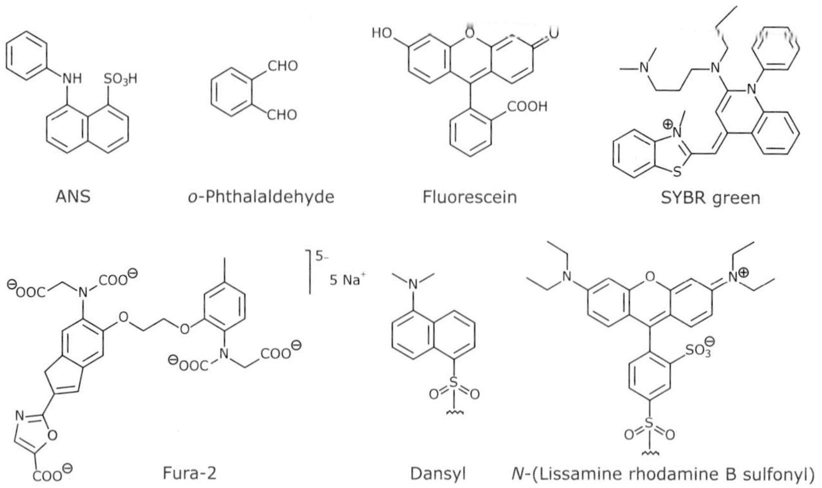

ANS *o*-Phthalaldehyde Fluorescein SYBR green

Fura-2 Dansyl *N*-(Lissamine rhodamine B sulfonyl)

Fig. 2.18. Structures of some extrinsic fluorophores. Fura-2 is a fluorescent chelator for divalent and higher valent metal ions (Ca^{2+}, Ba^{2+}, Sr^{2+}, Pb^{2+}, La^{3+}, Mn^{2+}, Ni^{2+} and Cd^{2+}).

not affect the mechanistic properties of the system under investigation. Secondly, its fluorescence emission needs to be sensitive to environmental conditions in order to enable monitoring of the molecular events. And lastly, the fluorophore must be tightly bound at a unique location.

Common non-conjugating extrinsic fluorophores for protein studies are 1-anilino-8-naphthalene sulfonate (ANS) and SYPRO Orange, which emit only weak fluorescence in a polar environment (i.e. in aqueous solution). However, in non-polar environments, for example when bound to hydrophobic patches on proteins, fluorescence emission of these dyes is significantly increased and the spectra show a hypsochromic shift; for example, for ANS, λ_{max} shifts from 475 to 450 nm. ANS is thus a valuable tool for assessment of the degree of non-polarity, and can also be used in competition assays to monitor binding of ligands and prosthetic groups. SYPRO Orange is the most frequently used dye in differential scanning fluorimetry (see Section 2.3.10).

Reagents such as fluorescamine, *o*-phthaladehyde and 6-aminoquinolyl-*N*-hydroxysuccinimidyl carbamate have been very popular conjugating agents used to derivatise amino acids for analysis. *o*-Phthalaldehyde, for example, is a non-fluorescent compound that reacts with primary amines (such as lysine side chains) and β-mercaptoethanol to yield a highly sensitive fluorophore.

Metal-chelating compounds with fluorescent properties are useful tools for a variety of assays, including monitoring of metal homeostasis in cells.

Chelators with fluorescent properties such as Fura-2 (see Fig. 2.18), Indo-1 and Quin-1 are widely used probes for calcium. However, as the chemistry of such compounds is based on metal chelation, cross-reactivity of the probes with other metal ions is possible.

The intrinsic fluorescence of nucleic acids is very weak and the required excitation wavelengths are too far in the UV region to be useful for practical applications. Numerous extrinsic fluorescent probes spontaneously bind to DNA and display enhanced emission. While in earlier days ethidium bromide was one of the most widely used dyes for this application, it has now generally been replaced by SYBR Green, as the latter probe poses fewer hazards for health and the environment and has no teratogenic properties like ethidium bromide. These probes bind DNA by intercalation of the planar aromatic ring systems between the base pairs of double-helical DNA. Their fluorescence emission in water is very weak and increases about 30-fold upon binding to DNA.

2.3.4 Quenching

In Section 2.3.1, we have seen that the quantum yield of a fluorophore is dependent on several internal and external factors. One of the external factors with practical implications is the presence of a quencher. A quencher molecule decreases the quantum yield of a fluorophore by non-radiating processes. The absorption (excitation) process of the fluorophore is not altered by the presence of a quencher. However, the energy of the excited state is transferred onto the quenching molecules. Two kinds of quenching process can be distinguished:

- dynamic quenching, which occurs by collision between the fluorophore in its excited state and the quencher; and
- static quenching, whereby the quencher forms a complex with the fluorophore. The complex has a different electronic structure compared with the fluorophore alone, and returns from the excited state to the ground state by non-radiating processes.

It follows intuitively that the efficacy of both processes is dependent on the concentration of quencher molecules. The mathematical treatment for each process is different, because of two different chemical mechanisms. Interestingly, in both cases the degree of quenching, expressed as I_0/I, is directly proportional to the quencher concentration. For collisional (dynamic) quenching, the resulting equation has been named the Stern–Volmer equation:

$$\frac{I_0}{I} - 1 = k_Q \times c(Q) \times \tau_0 \tag{2.13}$$

The Stern–Volmer equation relates the degree of quenching (expressed as I_0/I) to the molar concentration of the quencher $c(Q)$, the lifetime of the fluorophore τ_0 and the rate constant of the quenching process k_Q.

In the case of static quenching, I_0/I is related to the equilibrium constant K_a, which describes the formation of the complex between the excited fluorophore and the quencher, and the concentration of the quencher. Importantly, a plot of I_0/I versus $c(Q)$ yields for both quenching processes a linear graph with a y intercept of 1:

$$\frac{I_0}{I} - 1 = K_a \times c(Q) \qquad (2.14)$$

Thus, fluorescence data obtained by intensity measurements alone cannot distinguish between static or collisional quenching. The measurement of fluorescence lifetimes or the temperature/viscosity dependence of quenching can be used to determine the kind of quenching process. It should be added that both processes can also occur simultaneously in the same system.

The fact that static quenching is due to complex formation between the fluorophore and the quencher makes this phenomenon an attractive assay for binding of a ligand to a protein. In the simplest case, the fluorescence emission being monitored is the intrinsic fluorescence of the protein. While this is a very convenient titration assay when validated for an individual protein–ligand system, one has to be careful when testing unknown pairs, because the same decrease in intensity can occur by collisional quenching.

A highly effective quencher for fluorescence emission is oxygen, as well as the iodide ion. Usage of these quenchers allows surface mapping of biological macromolecules. For instance, iodide can be used to determine whether tryptophan residues are exposed to solvent.

2.3.5 Steady-state and time-resolved fluorescence

Steady-state fluorescence

Typical in-house fluorescence spectrophotometers are suitable for steady-state measurements only. As the recorded fluorescence spectrum of any sample is the superposition of all species present, one might have difficulties in interpretation if dynamic processes are going on within the sample. Therefore, the ideal samples for these experiments are equilibrated, static samples where no conversions of any species are taking place.

Time-resolved fluorescence

Elucidation of fluorescence in a time-resolved fashion is based on the lifetime of fluorophores. As we introduced in Section 2.3.1, the ability of a fluorophore to emit fluorescence is a time-dependent phenomenon. To enable fluorescence emission, a fluorophore has to be excited. The average time a molecule stays in the excited state is called the fluorescence lifetime.

The transition from the excited to the ground state can be treated as a first-order decay process, i.e. the number of molecules in the excited state decreases exponentially with time. If the number of molecules in the excited state is N_0, the number of molecules emitting fluorescence (and thereby returning to the ground state) can be calculated according to:

$$-\frac{dN}{dt} = k_f \times N_0, \tag{2.15}$$

which yields

$$N(t) = N_0 \times e^{-k_f t}. \tag{2.16}$$

The time it takes to reduce the number of fluorescence-emitting molecules to N_0/e is defined as the fluorescence lifetime τ_f:

$$\tau_f = \frac{1}{k_f} \tag{2.17}$$

By analogy to reaction kinetics, the exponential coefficient k_r is called the rate constant and is the reciprocal of the lifetime. The effective lifetime τ of excited molecules, however, differs from the fluorescence lifetime τ_f, as other, non-radiative processes also affect the number of molecules in the excited state (see Equation 2.12).

The emission of a single photon from a fluorophore follows a probability distribution. With time-correlated single-photon counting, the number of emitted photons can be recorded in a time-dependent manner following a pulsed excitation of the sample. By sampling the photon emission for a large number of excitations, the probability distribution can be constructed. The time-dependent decay of an individual fluorophore species follows an exponential distribution, and the time constant is thus termed the lifetime of this fluorophore. Curve fitting of fluorescence decays enables the identification of the number of species of fluorophores (within certain limits), and the calculation of the lifetimes for these species. In this context, different species can be different fluorophores or distinct conformations of the same fluorophore.

2.3.6 Fluorescence resonance energy transfer

To monitor the action and interactions of biological macromolecules, fluorescence resonance energy transfer (FRET) provides a very useful tool. The high specificity of the FRET signal allows monitoring of molecular interactions and conformational changes with high spatial (1–10 nm) and time (<1 ns) resolution. In particular, the possibility of localising and monitoring cellular structures and proteins in physiological environments makes this method very attractive (Brakmann & Nöbel, 2003).

FRET was first described by Förster in 1948 (Förster, 1948), and proven by Stryer and Haugland (Stryer & Haugland, 1967). The process can be explained in terms of quantum mechanics by a non-radiative energy transfer from a donor to an acceptor chromophore. The donor chromophore (D) is excited by light of the energy $h\nu_1$; energy is transferred by a non-radiative process to the acceptor chromophore (A), which then emits light of lesser energy $h\nu_2$ (see Fig. 2.19).

The requirements for this process are a reasonable overlap of emission and excitation spectra of donor and acceptor chromophores, close spatial vicinity of both chromophores (10–100 Å) and an almost parallel arrangement of their transition dipoles. The transfer rate k_T and the transfer

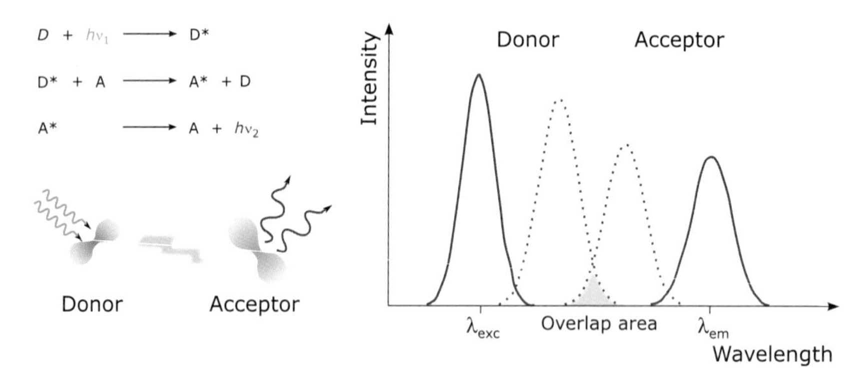

Fig. 2.19. Illustration of the fluorescence resonance energy transfer (FRET). Left: Pseudo-reaction scheme for the FRET processes between a donor (D) and acceptor (A) molecule. Right: Schematic fluorescence spectra of donor and acceptor molecules. The light required to excite the donor molecule is characterised by the fluorescence excitation spectrum (blue solid curve; λ_{exc}). The donor emission spectrum (red dotted curve) is not observed, but energy is transferred to the acceptor molecule due to the spectral overlap (orange) with the acceptor's fluorescence excitation spectrum (blue dotted curve). The acceptor molecule then emits light of higher wavelength ($\lambda_{em} > \lambda_{exc}$; red solid curve).

efficiency E_T depend on the distance R between donor and acceptor as per the following relationships:

$$k_T = \frac{1}{\tau_D}\left(\frac{R}{R_0}\right)^{-6} \tag{2.18}$$

$$E_T = \frac{k_T}{k_T + k_f^D + k_{IC}^D + k_{ISC}^D} = \frac{R_0^6}{R^6 + R_0^6} \tag{2.19}$$

τ_D is the fluorescence lifetime of the donor. The Förster radius R_0 is the critical distance for a donor–acceptor pair, where the probabilities of resonance transfer and intramolecular deactivation by radiative and non-radiative processes are equal. For a given donor–acceptor pair, R_0 is a constant with typical values between 10 and 50 Å.

The dependency of FRET on the distance between donor and acceptor makes this effect particularly suitable for biological applications, as the range of 10–100 Å is in the order of distances observed with biologically active macromolecules. In particular, if a process exhibits changes in molecular distances, FRET is considered in the choice of applicable approaches ('spectroscopic ruler').

FRET presents an ideal tool for monitoring biomolecular interactions and reactions, as effects can be observed even at low concentrations (as low as single molecules) and in different environments (different solvents, including living cells), and observation may be done in real time. This emphasises the variability of the method, and further development of fluorescence probes and their embedding will enhance the spectrum of possible applications.

FRET-based assays may be used to elucidate the effects of new substrates for different enzymes or putative agonists in a quick and quantitative manner. Furthermore, FRET detection can be used in high-throughput screenings, which makes it very attractive for drug development (Moshinsky *et al.*, 2003; Rogers *et al.*, 2012). However, applications of time-resolved FRET are in many cases limited to small peptide substrates.

Donor–acceptor pairs

In most cases, different extrinsic chromophores are used as donor and acceptor, which allows two possibilities to record FRET: either as donor-stimulated fluorescence emission of the acceptor or as fluorescence quenching of the donor by the acceptor. However, the same chromophore may be used as donor and acceptor simultaneously; in this case, the depolarisation of fluorescence (see Section 2.3.8) is the observed

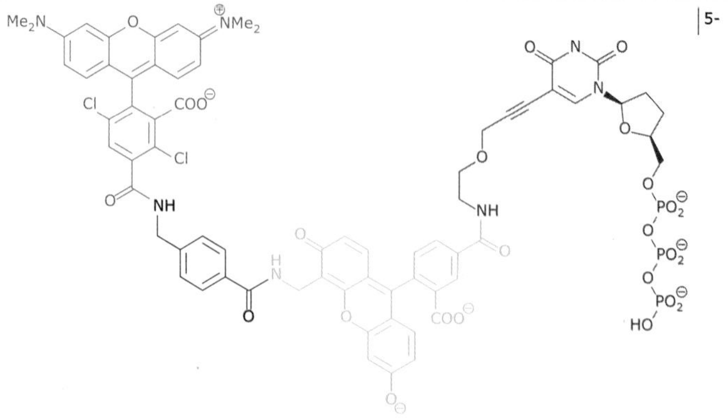

Fig. 2.20. The structure of one of the four BigDye™ terminators, ddT-EO-6CFB-dTMR, comprising: 5-carboxy-dichloro-rhodamine (magenta, FRET acceptor), 4-aminomethyl benzoate linker (black), 6-carboxy-4′-aminomethylfluorescein (green, FRET donor), a propargyl ethoxyamino linker (blue) and dUTP (red).

parameter. As non-FRET-stimulated fluorescence emission by the acceptor can result in undesirable background fluorescence, a common approach is usage of non-fluorescent acceptor chromophores.

BigDyes™ are widely used FRET fluorophores (Fig. 2.20). Since 1997, these fluorophores have generally been used as chain termination markers in automated DNA sequencing (Rosenblum *et al.*, 1997). As such, BigDyes™ are largely responsible for the high-throughput sequencing in genome projects.

The continuous development of chromophores and the steady increase in FRET applications are currently stimulating each other in a cyclic manner. Nowadays, FRET chromophores are available in a spectral range from 350 nm (ultraviolet) to 750 nm (IR). The IR chromophores are of particular interest due to their ability to be detected in living organisms with almost no background fluorescence. IR chromophores are characterised by high extinction coefficients, and they emit light of high intensity but very short fluorescence lifetime ($\tau_f < 4$ ns). Ideally, a donor with a long fluorescence lifetime and low quantum yield will be coupled with an acceptor that possesses a short lifetime but high quantum yield (Fig. 2.21).

Photochromic acceptors

A recent development is the design of photochromic acceptors for FRET applications (Fig. 2.22) (Giordano *et al.*, 2002; Song *et al.*, 2002). These acceptors become FRET acceptors only by light-induced isomerisation.

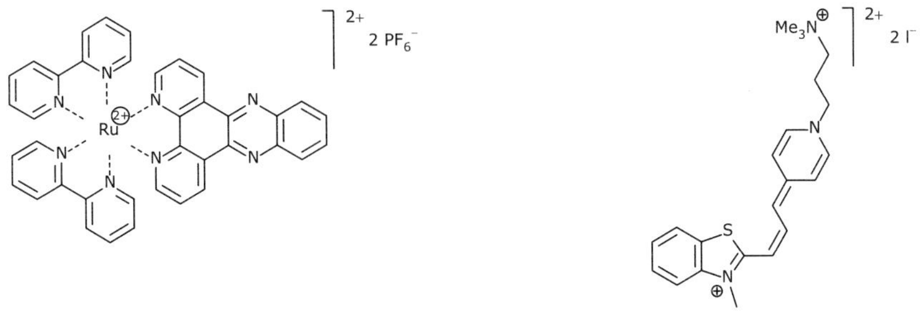

$[Ru(bpy)_2(dppz)]^{2+}$

BO-PRO-3

Fig. 2.21. IR chromophores used in FRET applications: $[Ru(bpy)_2(dppz)]^{2+}$, FRET donor, and BO-PRO-3, FRET acceptor. The use of luminescent transition metal complexes provides long fluorescence lifetimes (100 ns–10 μs) and an enhanced spectrum of absorption and emission wavelengths (Kang *et al.*, 2002). Quantum dots used in semiconductors like CdSe or CdTe might also be used as donors. The latter compounds possess tunable absorption spectra and sharp symmetrical emission spectra.

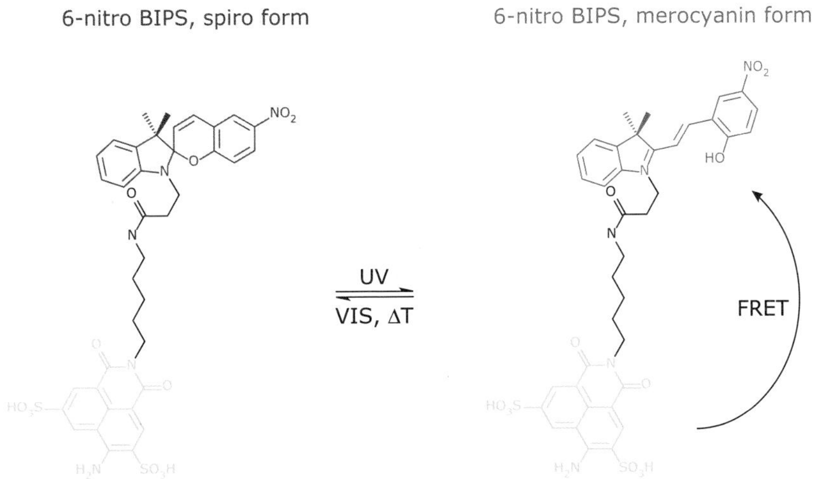

6-nitro BIPS, spiro form

6-nitro BIPS, merocyanin form

UV

VIS, ΔT

FRET

Fig. 2.22. Structure of photochromic FRET pair LYC–BIPS. The FRET donor LYC (green) is bound via a linker (black) to the photochromic acceptor 6-nitro BIPS (blue/magenta). Upon exposure to light in the near-UV region, the spiro form of the acceptor (blue) opens up to yield the merocyanin form (magenta), which acts as FRET acceptor for LYC.

Furthermore, they can be returned to their non-fluorescent forms by either thermal or photochemical means. The photochemical restoration requires a different wavelength from the one used for activation. Ideally, this activation/deactivation cycle can be switched reversibly for long time intervals. The FRET effect can then be measured repeatedly without photobleaching and with the donor fluorescence emission (no acceptor background noise!) as internal standard.

Bioluminescence resonance energy transfer

Bioluminescence resonance energy transfer (BRET) uses the FRET effect with native fluorescent or luminescent proteins as chromophores. The phenomenon is observed naturally for example with the sea pansy *Renilla reniformis*. It contains the enzyme luciferase, which oxidises luciferin (coelenterazin) with simultaneous emission of light at $\lambda = 480$ nm. This light directly excites green fluorescent protein (GFP), which, in turn, emits fluorescence at $\lambda = 509$ nm (Xu *et al.*, 1999; Boute *et al.*, 2002).

Fluorescent labelling of proteins with other proteins presents a useful approach to study various processes *in vivo*. Labelling can be done at the genetic level by generating fusion proteins. Monitoring of protein expression by GFP (Remington, 2011) is already an established technique, and further development of 'living colours' will lead to promising new tools.

While nucleic acids have been the main players in the genomic era, the post-genomic/proteomic era focuses on the gene products, the proteins. New proteins are being discovered and characterised, while others are already used within biotechnological processes. In particular, for classification and evaluation of enzymes and receptors, reaction systems can be designed such that the reaction of interest is detectable quantitatively using FRET donor and acceptor pairs.

For instance, detection methods for protease activity can be developed based on BRET applications. A protease substrate is fused to a GFP variant on the N-terminal side and dsRED on the C-terminal side. The latter protein is a red-fluorescing FRET acceptor and the GFP variant acts as a FRET donor. Once the substrate is cleaved by a protease, the FRET effect is abolished. This is used to directly monitor protease activity. With a combination of FRET analysis and two-photon excitation spectroscopy, it has also been possible to carry out a kinetic analysis (Kohl *et al.*, 2002).

A similar scheme is used to label human insulin receptor in order to assess quantitatively its activity. The insulin receptor is a glycoprotein with two α- and two β-subunits, which are linked by dithioether bridges. The binding of insulin induces a conformational change and causes a close spatial arrangement of both β-subunits. This, in turn, activates the tyrosine kinase activity of the receptor.

In pathological conditions like diabetes, the tyrosine kinase activity is different from that in healthy patients. Evidently, it is of great interest to find compounds that stimulate the same activity as insulin. By fusing the β-subunit of human insulin receptor to *Renilla reniformis* luciferase and yellow fluorescent protein (YFP), a FRET donor–acceptor pair is obtained, which reports the ligand-induced conformational change and precedes the signal transduction step. This reporter system is able to detect the effects of insulin and insulin-mimicking ligands in order to assess dose-dependent behaviour (Boute *et al.*, 2002).

Protein–protein and protein–DNA interactions

Whenever a change in the state of a protein (e.g. ligand binding, association with other proteins) is accompanied by a change in interatomic distances, the FRET effect is a phenomenon to be considered for observation of associated processes. In addition to colocalisation of proteins, FRET-based assays might therefore be used to monitor protein folding or conformational changes.

For instance, the three subunits of T4 DNA polymerase holoenzyme arrange around DNA in torus-like geometry. Using the tryptophan residue in one of the subunits as the FRET donor and a coumarin label at a cysteine residue in the adjacent subunit (FRET acceptor), the distance change between both subunits has been monitored and seven steps involved in opening and closing of the polymerase have been identified (Trakselis *et al.*, 2001). Other examples of this approach include studies of the architecture of *Escherichia coli* RNA polymerase (Heyduk & Niedziela-Majka, 2002), the calcium-dependent change of troponin (Kimura *et al.*, 2002) and structural studies of neuropeptide Y dimers (Bettio & Beck-Sickinger, 2001).

2.3.7 Fluorescence recovery after photobleaching (FRAP)

If a fluorophore is exposed to high-intensity radiation, it may be irreversibly damaged and lose its ability to emit fluorescence. Intentional bleaching of a fraction of fluorescently labelled molecules in a membrane can be used to monitor the motion of labelled molecules in certain (two-dimensional) compartments. Moreover, the time-dependent monitoring allows determination of the diffusion coefficient. A well-established application is the usage of phospholipids labelled with nitrobenzoxadiazole (NBD; e.g. NBD-phosphatidylethanolamine), which are incorporated into a biological or artificial membrane. The specimen is subjected to a pulse of high-intensity

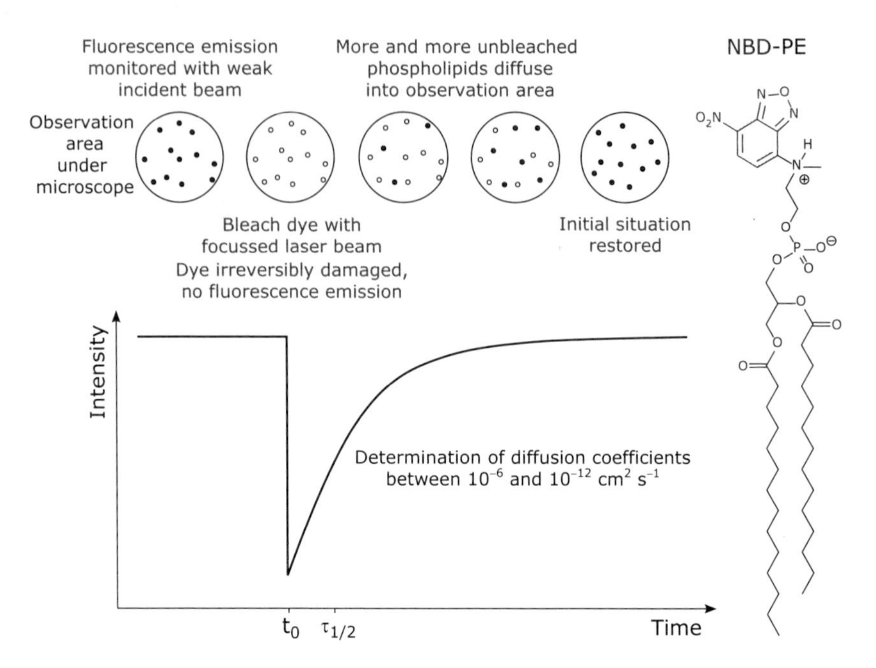

Fig. 2.23. Left: Schematic of a FRAP experiment. Time-based monitoring of fluorescence emission intensity enables determination of diffusion coefficients in membranes. Right: A commonly used fluorescent label in membrane FRAP experiments: chemical structure of the fluorophore NBD (red) conjugated to the lipid phosphatidylethanolamine (PE).

light (photobleaching), which causes a sharp drop in fluorescence in the observation area (Fig. 2.23). Re-emergence of fluorescence emission in this area is monitored as unbleached molecules diffuse into the observation area. From the time-dependent increase in fluorescence emission, the rate of diffusion of the phospholipid molecules can be calculated. Similarly, membrane proteins such as receptors or even proteins in a cell can be conjugated to fluorescent labels and their diffusion coefficients determined.

2.3.8 Fluorescence polarisation

A light source usually consists of a collection of randomly oriented emitters, and the emitted light is a collection of waves with all possible orientations of the $\vec{E}$ vectors (non-polarised light). Linearly polarised light is obtained by passing light through a polariser that transmits light with only a single plane of polarisation, i.e. it passes only those components of the $\vec{E}$ vector that are parallel to the axis of the polariser (Fig. 2.24). The intensity of transmitted light depends on the orientation of the polariser. Maximum transmission is achieved when the plane of polarisation is parallel to the

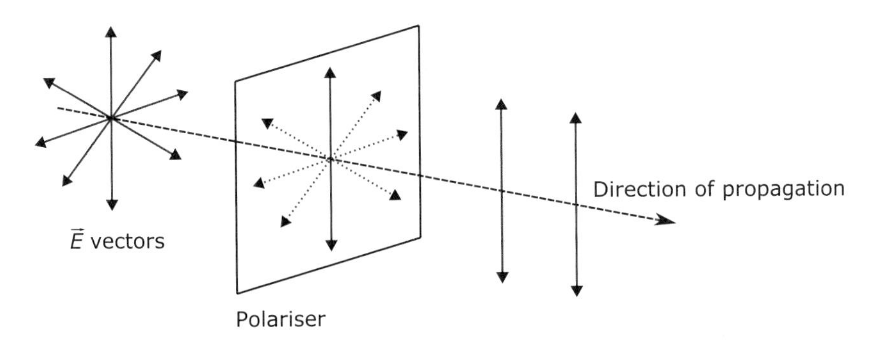

Fig. 2.24. Generation of linearly polarised light.

axis of the polariser; the transmission is zero when the orientation is perpendicular. The polarisation P and the anisotropy A are defined as

$$P = \frac{I_{\updownarrow} - I_{\leftrightarrow}}{I_{\updownarrow} + I_{\leftrightarrow}} \tag{2.20}$$

$$A = \frac{I_{\updownarrow} - I_{\leftrightarrow}}{I_{\updownarrow} + 2 \times I_{\leftrightarrow}} \tag{2.21}$$

$I_{\updownarrow}$ and $I_{\leftrightarrow}$ are the intensities observed parallel and perpendicular to an arbitrary axis. The polarisation can vary between -1 and $+1$; it is zero, when the light is unpolarised. Light with $0 < |P| < 0.5$ is called partially polarised. The definition for the anisotropy A considers all three directions for normalisation; hence, there is one term for the parallel direction (e.g. z direction) and two terms for the perpendicular directions (e.g. x and y directions).

Experimentally, this can be achieved in a fluorescence spectrometer by placing a polariser in the excitation path in order to excite the sample with polarised light. A second polariser is placed between the sample and the detector with its axis either parallel or perpendicular to the axis of the excitation polariser. The emitted light is either partially polarised or entirely unpolarised. This loss of polarisation is called fluorescence depolarisation.

Absorption of polarised light by a chromophore is highest when the plane of polarisation is parallel to the absorption dipole moment $\vec{\mu}_A$ (Fig. 2.25). More generally, the probability of absorption of exciting polarised light by a chromophore is proportional to $\cos^2\theta$, with θ being the angle between the direction of polarisation and the absorption dipole moment. Fluorescence emission, in contrast, does not depend on the absorption dipole moment but on the transition dipole moment $\vec{\mu}_E$. Usually, $\vec{\mu}_A$ and $\vec{\mu}_E$ are tilted relative to each other by about 10–40°. The probability of emission of polarised light at an angle ϕ with respect to

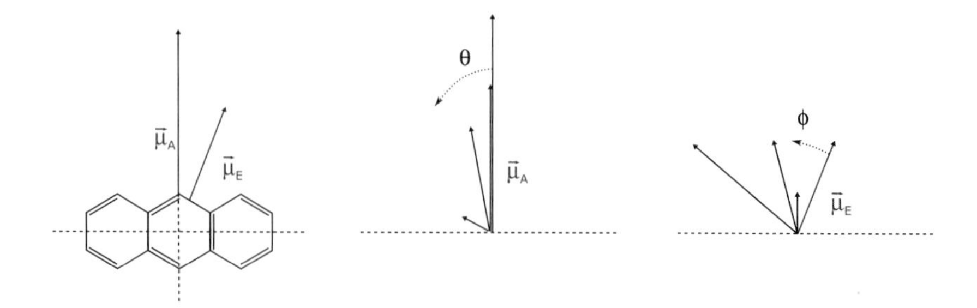

Fig. 2.25. Absorption dipole moment $\vec{\mu}_A$ (describing the probability of photon absorption) and transition dipole moment $\vec{\mu}_E$ (describing the probability for photon emission) for any chromophore are usually not parallel. Absorption of linearly polarised light varies with $\cos^2\theta$ and is at its maximum parallel to $\vec{\mu}_A$. Emission of linearly polarised light varies with $\sin^2\phi$ and is highest at a perpendicular orientation to $\vec{\mu}_E$.

the transition dipole moment is proportional to $\sin^2\phi$ and thus at its maximum in a perpendicular orientation.

As a result, if the chromophores are randomly oriented in solution, the polarisation P is less than 0.5. It is thus evident that any process that leads to a deviation from random orientation will give rise to a change of polarisation. This is certainly the case when a chromophore becomes more static. Furthermore, one needs to consider Brownian motion. If the chromophore is a small molecule in solution, it will be rotating very rapidly. Any change in this motion due to temperature changes, changes in viscosity of the solvent or binding to a larger molecule will therefore result in a change of polarisation.

Analysis of fluorescence polarisation allows conclusions to be drawn as to:

- rotation mobility (rotation correlation time);
- orientation of fluorophores in a matrix; and
- viscosity of the environment surrounding the fluorophore.

Typically, such experiments are carried out as time-resolved fluorescence measurements with the oscillating emission polariser oriented first perpendicular and then parallel to the linearly polarised excitation beam; detection may be either through a single-photon counter with oscillating polariser or a beam splitter distributing the light to two counters with fixed polarisers. The intrinsic anisotropy of a fluorophore, A_0, is usually measured in a matrix of frozen polyol. In solution, the fluorophores are free to move and the anisotropy A decreases exponentially with time; the rotation correlation time can thus be determined from the power of the fitted exponential function. The ratio of the intrinsic anisotropy of a fluorophore and its anisotropy in solution depends on how quickly the molecules can change their orientation (rotational lifetime ϕ) compared with the fluorescence lifetime (τ):

$$\frac{A_0}{A} = 1 + \frac{\tau}{\phi} \tag{2.22}$$

2.3.9 Fluorescence cross-correlation spectroscopy

With fluorescence cross-correlation spectroscopy, the temporal fluorescence fluctuations between two differently labelled molecules can be measured as they diffuse through a small sample volume. Cross-correlation analysis of the fluorescence signals from separate detection channels extracts information of the dynamics of the dual-labelled molecules. Fluorescence cross-correlation spectroscopy has thus become an essential tool for the characterisation of diffusion coefficients, binding constants and kinetic rates of binding, and for determining molecular interactions in solutions and cells.

2.3.10 Differential scanning fluorimetry (DSF)

Differential scanning fluorimetry (ThermoFluor) is a popular method that exploits the difference in protein folding stability in the absence and presence of a bound ligand. The protein target is subjected to thermal denaturation and the unfolding process is monitored by an amphiphilic fluorescence dye (see Extrinsic fluorescence, above), typically SYPRO Orange (Fig. 2.26). Amphiphilic dyes (such as SYPRO Orange, ANS and others) exhibit a strong fluorescence enhancement when they are less exposed to hydrophilic solvent.

In the folded state of a protein, the contact of hydrophobic side chains with hydrophilic solvent is minimised, and the amphiphilic dye thus exhibits low fluorescence emission (Fig. 2.27). A gradual increase in the temperature leads to unfolding of the protein and more hydrophobic areas become exposed, which are reported by the fluorescence dye through increased emission. In many cases, aggregation effects are observed once protein molecules are unfolded, and this can lead to shielding of hydrophobic areas on individual protein molecules; thus, a decrease in fluorescence emission is frequently observed at high temperatures in the DSF experiment.

Differential scanning fluorimetry is a rapid and inexpensive screening method to identify ligands that bind and stabilise purified proteins. The 'melting temperature' ($T_{1/2}$ or T_m) is measured by an increase in the

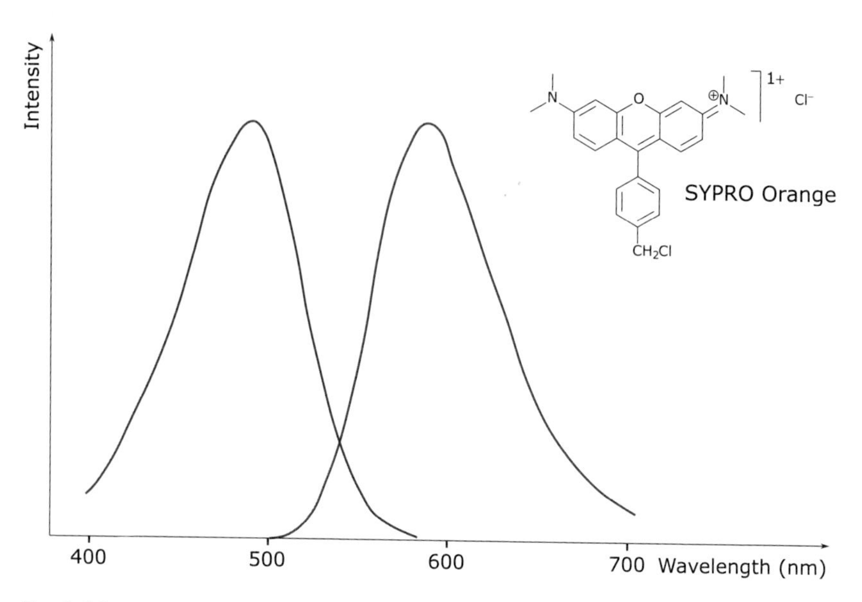

Fig. 2.26. Fluorescence excitation and emission spectra of SYPRO Orange, a dye frequently used in DSF.

fluorescence of the amphiphilic fluorescence dye. The difference in the melting temperatures of a protein solution in the absence and presence of a ligand, $\Delta T_{1/2}$, is related to the binding affinity of the small molecule, which can be a low-molecular-mass compound, a peptide or a nucleic acid. Experiments are typically performed in real-time PCR instruments with multi-well plates, thus offering a relatively inexpensive medium-throughput screening of ligand libraries.

A few caveats exist, however. In some cases, entropic effects may hinder straightforward interpretation. Secondly, not all proteins may be amenable to this methodology, as the amphiphilic dye may bind to hydrophobic areas on the protein exposed in the native state. In this case, the recorded fluorescence emission of the reporter dye shows little or no change in intensity as the thermal denaturation progresses. Lastly, some metal ions (e.g. Co^{2+}, Hg^{2+}) can interfere with fluorescence emission from the reporter dye.

2.3.11 Fluorescence microscopy and high-throughput assays

Fluorescence emission as a means of monitoring is a valuable tool for many biological and biochemical applications. We have already seen the usage of fluorescence monitoring in DNA sequencing; the technique is inseparably tied in with the success of projects such as genome deciphering.

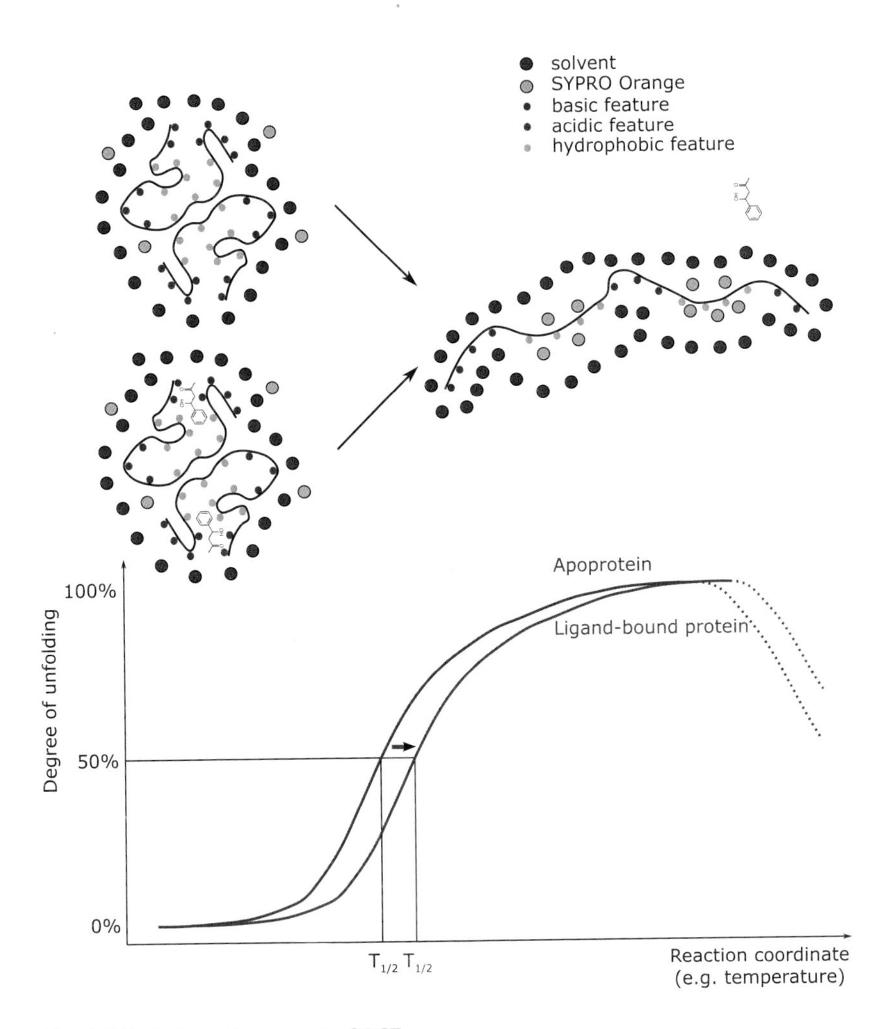

Fig. 2.27. Schematic concept of DSF.

Fluorescence techniques are also indispensable methods for cell biological applications with fluorescence microscopy. Proteins (or biological macromolecules) of interest can be tagged with a fluorescent label such as GFP from the jelly fish *Aequorea victoria* or red fluorescent protein from *Discosoma striata*, if spatial and temporal tracking of the tagged protein is desired. Alternatively, the use of GFP spectral variants such as cyan fluorescent protein as a fluorescence donor and YFP as an acceptor allows investigation of mechanistic questions by using the FRET phenomenon. Specimens with cells expressing the labelled proteins are illuminated with light of the excitation wavelength and then observed through a filter that excludes the exciting light and only transmits the fluorescence emission.

The recorded fluorescence emission can be overlaid with a visual image computationally, and the composite image then allows localisation of the labelled species. If different fluorescent labels with distinct emission wavelengths are used simultaneously, even colocalisation studies can be performed.

2.4 Luminometry

In the preceding section, we already mentioned the method of BRET and its main workhorse, luciferase. Generally, fluorescence phenomena depend on the input of energy in the form of electromagnetic radiation. However, emission of electromagnetic radiation from a system can also be achieved by prior excitation in the course of a chemical or enzymatic reaction. Such processes are summarised as luminescence.

2.4.1 Principles

Luminometry is the technique used to measure luminescence, which is the emission of electromagnetic radiation in the energy range of visible light as a result of a reaction. Chemiluminescence arises from the relaxation of excited electrons transitioning back to the ground state. The prior excitation occurs through a chemical reaction that yields a fluorescent product. For instance, the reaction of luminol with oxygen produces 3-aminophthalate, which possesses a fluorescence spectrum that is then observed as chemiluminescence. In other words, the chemiluminescence spectrum is the same as the fluorescence spectrum of the product of the chemical reaction.

Bioluminescence describes the same phenomenon, only that the reaction leading to a fluorescent product is an enzymatic reaction. The most commonly used enzyme in this context is certainly luciferase. The light is emitted by an intermediate complex of luciferase with the substrate ('photoprotein'). The colour of the light emitted depends on the source of the enzyme and varies between 560 nm (greenish yellow) and 620 nm (red). Bioluminescence is a highly sensitive method, due to the high quantum yield of the underlying reaction. Some luciferase systems work with almost 100% efficiency. For comparison, the incandescent light bulb loses about 90% of the input energy to heat.

Because luminescence does not depend on any optical excitation, problems with autofluorescence in assays are eliminated.

2.4.2 Instrumentation

As no electromagnetic radiation is required as a source of energy for excitation, no light source or monochromator is required. Luminometry can be performed with a rather simple setup, where a reaction is started in a cuvette or mixing chamber and the resulting light is detected by a photometer. In most cases, a photomultiplier tube is needed to amplify the output signal prior to recording. Also, it is fairly important to maintain a strict temperature control, as all chemical, and especially enzymatic, reactions are sensitive to temperature.

2.4.3 Applications

Chemiluminescence

Luminol and its derivatives can undergo chemiluminescent reactions with high efficiency. For instance, enzymatically generated H_2O_2 may be detected by the emission of light at 430 nm wavelength in the presence of luminol and a peroxidase such as microperoxidase, a haem-containing peptide from cytochrome *c*.

Competitive binding assays may be used to determine low concentrations of hormones, drugs and metabolites in biological fluids. These assays depend on the ability of proteins such as antibodies and cell receptors to bind specific ligands with high affinity. Competition between labelled and unlabelled ligand for appropriate sites on the protein occurs. If the concentration of the protein, i.e. the number of available binding sites, is known, and a limited but known concentration of labelled ligand is introduced, the concentration of unlabelled ligand can be determined under saturation conditions when all sites are occupied. Exclusive use of labelled ligand allows the determination of the concentration of the protein and thus the number of available binding sites.

Molecular oxygen in its singlet state (see Sections 1.1.2 and 2.4.4) also exhibits chemiluminescence, and is thus amenable to this technique. The effects of pharmacological and toxicological agents on leukocytes and other phagocytic cells can be studied by monitoring luminescence arising from nascent O_2 phagocytosis.

Bioluminescence

Firefly luciferase is mainly used to measure ATP concentrations. The bioluminescence assay is rapidly carried out with accuracies comparable to spectrophotometric and fluorimetric assays. However, with a detection limit of 10^{-15} M, and a linear range of 10^{-12}–10^{-6} M ATP, the luciferase assay is superior to the photometric assays in terms of sensitivity. Generally, all enzymes and metabolites involved in ATP interconversion reactions may be assayed in this method, including ADP, AMP, cyclic AMP and the enzymes pyruvate kinase, adenylate kinase, phosphodiesterase, creatine kinase, hexokinas, and ATP sulfurase. Other substrates include creatine phosphate, glucose, GTP, phosphoenolpyruvate and 1,3-diphosphoglycerate.

The main application of bacterial luciferase is the determination of electron transfer cofactors, such as nicotinamide adenine dinucleotides (and phosphates) and flavin mononucleotides in their reduced states, for example NADH, NADPH and $FMNH_2$. Similar to the firefly luciferase assays, this method can be applied to a whole range of coupled RedOx enzyme reaction systems. The enzymatic assays are again much more sensitive than the corresponding spectrophotometric and fluorimetric assays, and a concentration range of 10^{-9}–10^{-12} M can be achieved. The NADPH assay is less sensitive than the NADH assay by a factor of 20.

2.4.4 Amplified luminescent proximity homogeneous assay

A chemiluminescence method originally called luminescent oxygen channelling immunoassay (LOCI) was reported in 1994 (Ullman *et al.*, 1994) and has since been further developed and marketed as the amplified luminescent proximity homogeneous assay (ALPHA). The method is based on two types of microbeads, donor and acceptor, which are coated with a hydrogel to minimise non-specific binding and self-aggregation. Conjugation of biomolecules to the beads is possible by reactive aldehyde groups on the bead surface. The geometry of the beads has been designed such that they do not sediment in buffers, thus making the assay suitable for automated liquid handling and high-throughput screening. However, their size does allow separation by centrifugation or filtration.

The donor beads contain phthalocyanine, a photosensitiser, that converts oxygen from its native triplet state ($^3O_2$; total spin $S=1$) to the excited singlet state ($^1O_2$; total spin $S=0$) when illuminated with light of 680 nm wavelength (Fig. 2.28). The acceptor beads contain thiophene derivatives. Within the limited lifetime (half-life 4 μs), $^1O_2$ can diffuse about 0.2 μm in solution. If an acceptor bead is within this radius, the energy can be transferred from the excited singlet oxygen to the thiophene acceptor, which will then emit light in the range of 520–620 nm. Otherwise, the oxygen molecule transitions back to its triplet ground state without any radiation being emitted. The requirement for spatial proximity in this assay

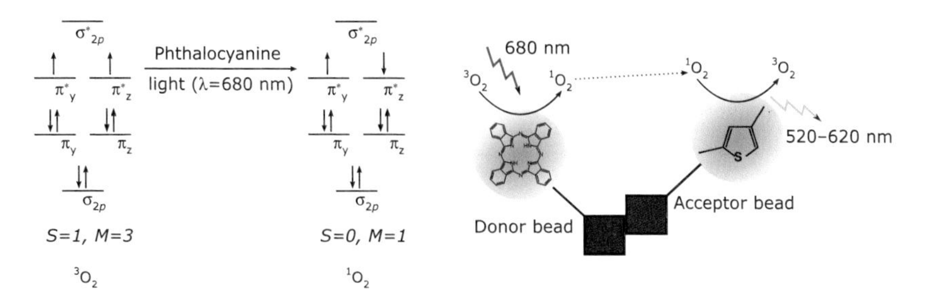

Fig. 2.28. Principle of the ALPHA technique.

in order for light emission to occur provides a generic detection method that allows a variety of biochemical assays to be monitored.

For example, the possible interaction of two proteins can conveniently be monitored using this technology. One protein needs to be coupled to the donor beads and the other to the acceptor beads. If a direct interaction occurs, energy transfer from singlet oxygen generated on the acceptor beads can be transferred to the thiophene moiety on the acceptor beads, as both beads are held in a spatial vicinity through the interaction between the two proteins. A variety of assay types has been developed in addition to basic protein–protein interaction monitoring, and includes cGMP detection and kinase substrate detection, as well as kinase activity assays. ALPHA screens enjoy great popularity as they can easily be transferred to high-throughput screening applications.

2.5 Circular dichroism spectroscopy

In Section 2.3.8, we have already seen that electromagnetic radiation oscillates in all possible directions and that it is possible to preferentially select waves oscillating in a single plane, as required for fluorescence polarisation studies. The phenomenon of polarised light was described as early as 1678 by Huygens based on investigations of a crystalline form of $CaCO_3$ called Iceland spar. About 130 years later, in the course of investigating the properties of polarised light, Biot and others discovered the optical rotation of many organic compounds and sugar solutions. Of particular importance in this context was the phenomenon called mutarotation, which describes the interconversion of the α- and β-anomers of cyclic sugars. The optical rotatory power – the ability of a compound to change the angle of polarisation of plane-polarised light – became manifest as a special property of optically active isomers, allowing the rotation of linearly polarised (or plane-polarised) light. Optically active isomers are compounds of identical chemical composition and topology but whose mirror images cannot be superimposed; such compounds are called chiral.

2.5.1 Linearly and circularly polarised light

Light is electromagnetic radiation where the electric vector ($\vec{E}$) and the magnetic vector ($\vec{M}$) are perpendicular to each other. Each vector undergoes an oscillation as the light travels along the direction of propagation, resulting in a sine-like waveform of each, the $\vec{E}$ and $\vec{M}$ vectors. A light source usually consists of a collection of randomly oriented emitters. Therefore, the emitted light is a collection of waves with all possible orientations of the $\vec{E}$ vectors. This light is non-polarised. Linearly or plane-polarised light is obtained by passing light through a polariser that transmits light with only a single plane of polarisation, i.e. it passes only those components of the $\vec{E}$ vector that are parallel to the axis of the polariser (see Fig. 2.24). If the $\vec{E}$ vectors of two electromagnetic waves are 1/4 wavelength out of phase and perpendicular to each other, the vector that is the sum of the $\vec{E}$ vectors of the two components rotates around the direction of propagation so that its tip follows a helical path. Such light is called circularly polarised (Fig. 2.29).

While the $\vec{E}$ vector of circularly polarised light always has the same magnitude but a varying direction, the direction of the $\vec{E}$ vector of linearly polarised light is constant; it is its magnitude that varies. Alternatively,

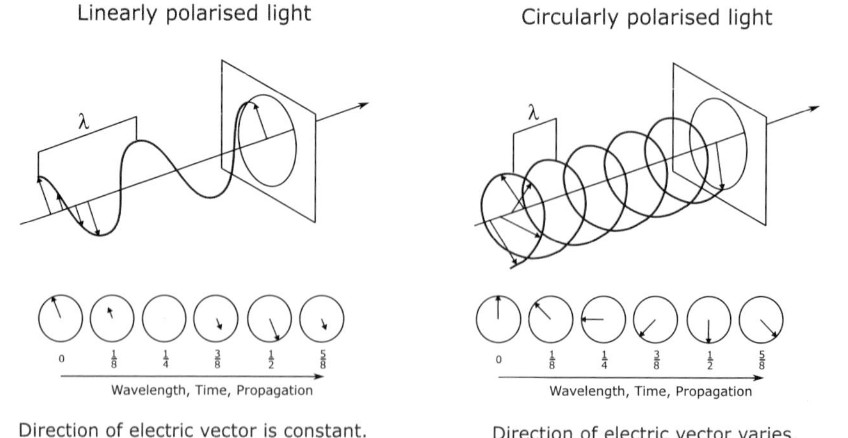

Fig. 2.29. Linearly (plane) and circularly polarised light.

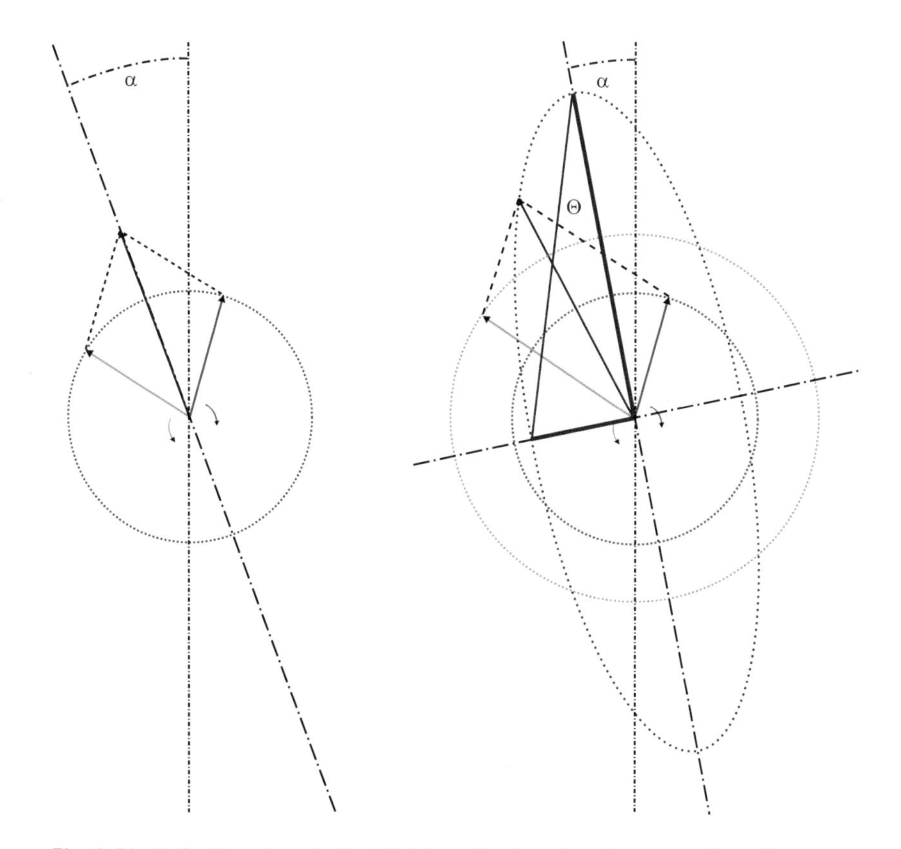

Fig. 2.30. Left: Linearly polarised light can be thought of as consisting of two circularly polarised components with opposite 'handedness'. The vector sum of the left-handed (yellow) and right-handed (green) circularly polarised light yields linearly polarised light. Right: If the amplitudes of left- and right-handed polarised components differ, the resulting light is elliptically polarised. The composite vector (red) will trace the ellipse shown in blue. The ellipse is characterised by a major and a minor axis. The ratio of these two axes yields $\tan\theta$. θ is the ellipticity.

with the help of vector algebra, one can think of linearly polarised light as a composite of two circularly polarised beams with opposite handedness (Fig. 2.30).

2.5.2 Polarimetry and optical rotation dispersion

When an optically active sample is transilluminated with linearly polarised light, the velocities of both circular components of polarisation differ after passing the sample. This effect is called optical rotatory dispersion (ORD).

Polarimetry essentially measures the angle through which the plane of polarisation is changed after linearly polarised light is passed through a solution containing a chiral substance. Optical rotatory dispersion spectroscopy is a technique that measures this ability of a chiral substance to change the plane polarisation as a function of the wavelength. The angle α_λ between the plane of the resulting linearly polarised light and that of the incident light is dependent on the refractive index for left (n_{left}) and right (n_{right}) circularly polarised light. The refractive index can be calculated as the ratio of the speed of light *in vacuo* and the speed of light in matter:

$$\alpha_\lambda = \frac{180^\circ \times d}{\lambda} \times (n_{left} - n_{right}), \tag{2.23}$$

with the refractive index n being defined as:

$$n = \frac{c}{c_{matter}}, \tag{2.24}$$

where c is the speed of light *in vacuo*, c_{matter} the speed of light in matter, d the thickness of the sample and λ the wavelength. The unit of α_λ is $[\alpha_\lambda] = 1^\circ$.

After normalisation against the amount of substance present in the sample (thickness of sample/cuvette length d, and mass concentration ρ^*), a substance-specific constant $[\alpha]_\lambda$ is obtained that can be used to characterise chiral compounds:

$$[\alpha]_\lambda = \frac{\alpha_\lambda}{\rho^* \times d}, \tag{2.25}$$

with $[\alpha_\lambda]$ having the unit $[[\alpha]_\lambda] = 1^\circ\ cm^2\ g^{-1}$, if the mass concentration is provided in the unit of $[\rho^*] = 1\ g\ ml^{-1}$, and the path length is provided in $[d] = 1$ cm.

2.5.3 Circular dichroism

In addition to changing the plane of polarisation, an optically active sample also shows unusual absorption behaviour. Left- and right-handed polarised components of the incident light are absorbed differently by the sample (absorbances A_{left} and A_{right}), which yields a difference in the absorption coefficients $\Delta\varepsilon = \varepsilon_{left} - \varepsilon_{right}$. This latter difference is called circular dichroism (CD). The difference in absorption coefficients $\Delta\varepsilon$ (i.e. CD) is the observed quantity in CD experiments:

$$\Delta A = A_{left} - A_{right} \Rightarrow \Delta\varepsilon = \frac{\Delta A}{\rho^* \times d}, \tag{2.26}$$

with $[\Delta\varepsilon] = 1\ cm^2\ g^{-1}$, if the mass concentration is provided in the dimension of $[\rho^*] = 1\ g\ ml^{-1}$, and the path length is provided in $[d] = 1$ cm. Alternatively, if using the molar concentration c instead of the mass concentration ρ^*:

$$\Delta\varepsilon = \frac{\Delta A}{c \times d}, \tag{2.27}$$

the CD can be reported in units of $[\Delta\varepsilon] = 1\,\text{l mol}^{-1}\,\text{cm}^{-1}$.

The CD signal is a very small difference between two large original parameters, which poses technical challenges to cope with poor signal-to-noise ratios. With proteins, for example, $\Delta\varepsilon$ is typically less than $10\,\text{l mol}^{-1}\,\text{cm}^{-1}$, and the CD is therefore only 1/1000th of the molar absorption coefficient of the sample.

ΔA represents the difference in absorption and $\Delta\varepsilon$ the difference in absorption coefficients; $\Delta\varepsilon$ is also the observed quantity in CD experiments. Historically, results from CD experiments are reported as ellipticity (Fig. 2.30), which is defined as:

$$\Delta\Theta_\lambda = \ln(10) \times \frac{180^\circ}{4\pi} \times \Delta\varepsilon \times c \times d, \tag{2.28}$$

with the unit $[\Delta\Theta_\lambda] = 1^\circ$.

Normalisation of Θ_λ similar to the ORD yields the molar ellipticity:

$$\Delta\theta_\lambda = \frac{M \times \Theta_\lambda}{10 \times \rho^* \times d} = \frac{\ln(10)}{10} \times \frac{180^\circ}{2\pi} \times \Delta\varepsilon \approx 3.298 \times \Delta\varepsilon, \tag{2.29}$$

with the unit $[\Delta\theta_\lambda] = 1^\circ\,\text{cm}^2\,\text{mol}^{-1}$.

It is common practice to display graphs of CD spectra with the molar ellipticity in the unit of $1^\circ\,\text{cm}^2\,\text{dmol}^{-1} = 10^\circ\,\text{cm}^2\,\text{mol}^{-1}$ on the ordinate axis.

Three important conclusions can be drawn:

- ORD and CD are the manifestation of the same underlying phenomenon.
- If an optically active molecule has a positive CD, then its enantiomer will have a negative CD of exactly the same magnitude.
- The phenomenon of CD can only be observed at wavelengths where the optically active molecule has an absorption band.

2.5.4 The chromophores of protein secondary structure

In Section 2.2.2, we saw that the peptide bond in proteins possesses UV absorption bands in the area of 220–190 nm, due to $\pi \rightarrow \pi^*$ transitions. This wavelength range is part of the so-called vacuum UV (VUV), because the strong absorption of oxygen renders these wavelengths inaccessible and requires the use of a vacuum. In practice, however, the light path in CD spectrometers is filled with nitrogen gas to enable measurement at wavelengths down to 190 nm. The carbon atom vicinal to the peptide bond (the Cα atom) is asymmetric and a chiral centre in all amino acids except glycine. This chirality induces asymmetry into the peptide bond chromophore. Because of the serial arrangement of the peptide bonds making up the backbone of a protein, the individual chromophores couple with each other. The (secondary) structure of a polypeptide thus induces an 'overall chirality', which gives rise to the CD phenomenon of a protein in the wavelength interval 260–190 nm.

With protein CD, the molar ellipticity θ is often reported as mean residue ellipticity θ_{res}, owing to the fact that the chromophores responsible for the chiral absorption phenomenon are the peptide bonds. Therefore, the number of chromophores of a polypeptide in this context is equal to the number of residues. Because of the law of Lambert–Beer, the number of chromophores is proportional to the magnitude of absorption, i.e. in order to normalise the spectrum of an individual polypeptide for reasons of comparison, the CD has to be scaled by the number of peptide bonds:

$$\theta_{res} = \frac{\theta_{\lambda}}{N}. \tag{2.30}$$

Protein CD spectroscopy is carried out either as far-UV CD (260–190 nm; preferably 260–170 nm, but this requires synchrotron sources) or near-UV CD (300–250 nm). For secondary structure analysis, the wavelength region of the far-UV is required, as it assesses the CD of the peptide (amide) bonds, which are responsible for the secondary structure of the protein backbone. Protein CD in the near-UV CD is due to aromatic amino acid side chains (phenylalanine, tyrosine and tryptophan), as well as cysteine and disulfide bonds; absorption, dipole orientation and the immediate environment of these side chains give rise to CD signals in the near-UV.

2.5.5 Visible and vibrational CD

As we have established earlier, CD can only be observed at wavelengths where absorption of light occurs. All molecules possess vibrational modes and show thus absorption of light in the IR region (see Section 2.7). If the vibration occurs at a bond involving a chiral centre, it is possible to detect the CD at the wavelength characterising that vibrational mode. In this case, vibrational CD spectroscopy can be employed; it is the extension of CD spectroscopy into wavelength regions of IR and near-IR (NIR) light. It has become very popular for characterisation of organic small molecules such as natural products.

Similarly, light with wavelengths in the visible range can be employed for CD spectroscopy. This approach finds frequent applications in the characterisation of transition metals, as individual *d-d* electronic transitions can be resolved as separate bands. For example, if protein-bound metal ions are coordinated in a chiral environment, visible CD spectra can be used to characterise the metal–protein interactions, as free metal ions in solution (uncoordinated and thus not in a chiral environment) do not possess a CD signal. As only protein-bound metals contribute to the CD signal, binding stoichiometries and effects of external factors such as pH can be investigated. Examples of proteins that have been investigated with this method include the prion protein, a copper-binding cell-surface glycoprotein. A set of empirical rules for predicting the appearance of visible CD spectra for square-planar complexes involving histidine and main-chain coordination has been established based on these studies (Klewpatinond & Viles, 2007).

Metal coordination complexes with organic or inorganic ligands can also be subjected to visible CD spectroscopy. Optical activity in transition metal-ion complexes has been attributed to the following effects:

- A configuration effect is due to the chiral arrangement of chelate rings around the metal centre. The inherent dissymmetry is the equivalent in coordination compounds of the chiral carbon atoms in organic compounds.
- A vicinal effect arises from an optically active ligand if its chirality is induced into electronic transitions involving mainly the metal centre.
- A conformational effect may be present if the ligand preferentially assumes a chiral conformation (called a λ- or δ-enantiomer).
- A symmetry distortion effect may occur if chiral distortions exist in the coordination compound.

In a particular coordination complex, several of these effects may be operating simultaneously.

Due to the hydroxyl and acetal groups, carbohydrates possess electronic transitions below 190 nm, which is the short-wavelength limit of conventional CD spectrophotometers in aqueous solution. For spectra at wavelengths significantly below 190 nm, the wavelength range of VUV (100–200 nm) can be employed; however, this requires synchrotron radiation. The CD of molecules with unsubstituted hydroxyl groups such as carbohydrates can thus only be investigated using VUV-CD.

2.5.6 Instrumentation

The basic layout of a CD spectrometer follows that of a single beam UV absorption spectrometer. However, owing to the nature of the measured effects, an electro-optic modulator, as well as a more sophisticated detector, is needed.

Generally, left and right circularly polarised light is passing through the sample in an alternating fashion. This is achieved by an electro-optic modulator, which is a crystal that transmits either the left- or right-handed polarised component of linearly polarised light, depending on the polarity of the electric field that is applied by alternating currents. The photomultiplier detector produces a voltage proportional to the ellipticity of the resultant beam emerging from the sample. The light source of the spectrometer is continuously flushed with nitrogen to avoid the formation of ozone and help to maintain the lamp. The replacement of air in the instrument light path with pure nitrogen further helps to suppress unwanted absorption of light at the high end of the UV region by oxygen.

CD spectrometry involves measuring a very small difference between two absorption values that are large signals. The technique is thus very susceptible to noise, and measurements must be carried out carefully. Some practical considerations involve having a clean quartz cuvette and using buffers with low concentrations of additives. While this is sometimes tricky with protein samples, reducing the salt concentrations to values as low as 5 mM helps to obtain good spectra. Also, filtered solutions should be used to avoid any turbidity of the sample that could produce scatter. Saturation of the detector, which is becoming more critical with lower wavelengths, must be avoided. Therefore, good spectra are only obtained with a relatively narrow range of protein concentrations where enough sample is present to produce a good signal but does not saturate the detector. Typical protein concentrations are 0.03–0.3 mg ml^{-1} for a 0.1 mm pathlength.

In order to calculate specific ellipticities (mean residue ellipticities) and to be able to compare the CD spectra of different samples with each other, the concentration of the sample must be known accurately. Provided the protein possesses sufficient amounts of UV/Vis-absorbing chromophores, it is thus advisable to subject the CD sample to a protein concentration determination by UV/Vis as described in Section 2.2.8.

As the CD signal in an isotropic solution is extremely small, this requires accumulating measurements to achieve meaningful spectra. Spectrum acquisition can thus take in the order of 30 min to several hours. In a recent development of the vibrational CD technique (Rhee *et al.*, 2009), short pulses of linearly polarised IR light are used to excite vibrations of the chiral sample. The resulting oscillating dipoles perform a free induction decay (as in nuclear magnetic resonance techniques; see Section 3.2.1) and emit their chiral information encoded in the amplitude and phase of the generated electric field. By analysing only the signal that is polarised orthogonal to the incident pulses, a weak chiral signal can be isolated from the achiral background. In order to amplify the weak signal, an IR reference pulse with higher intensity than the signal is overlayed onto the free induction decay, which gives rise to an interference pattern that can be deconvoluted by Fourier transformation. The further development of this technique may in the near future allow time-resolved CD spectroscopy – a feature that is not possible with current instruments.

2.5.7 Applications

Circular dichroism is a useful tool in the determination of the absolute configuration of small molecules. As light absorption is associated with energetic transitions, all molecules containing stereogenic centres, chiral axes and helical structures in proximity to a conjugated π system or a metal complex will show an induced CD. As all of these are structural features of a molecule, quantum chemical calculations of the CD spectra can be used to determine the absolute configuration.

The main application for protein CD spectroscopy is the verification of the adopted secondary structure (Fig. 2.31 and Table 2.13); this analysis is based on the chromophores for secondary structure (see Section 2.5.4). The application of CD to determine the tertiary structure is limited, owing to the inadequate theoretical understanding of the effects of different parts of the molecules at this level of structure.

Rather than analysing the secondary structure of a 'static sample', different conditions can be tested. For instance, some peptides adopt different secondary structures when in solution or membrane bound. The

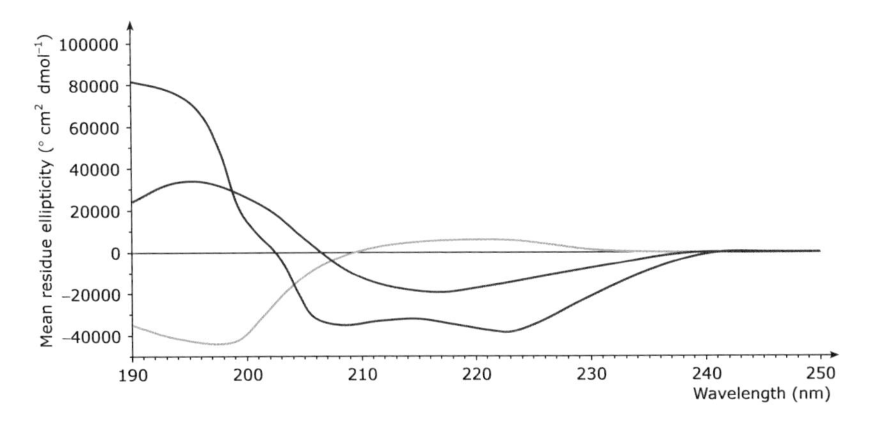

Fig. 2.31. Circular dichroism spectra for three standard secondary structures according to the method of Fasman and colleagues (Perczel *et al.*, 1992). An α-helical peptide is shown in blue, a peptide adopting β-strand structure in red and a random-coil peptide in yellow.

Table 2.13. Examples of extrema occurring in CD spectra of different secondary structure elements (approximate values).

Secondary structure element	Minimum/maximum		
	Wavelength (nm)	θ_{res} (° cm^2 mol^{-1})	θ_{res} (° cm^2 dmol^{-1})
α-Helix	222	-3×10^5	−30 000
	207	-3×10^5	−30 000
	192	6×10^5	60 000
β-Strand	218	-1.5×10^5	−15 000
	195	2×10^5	20 000
Random coil	220	1×10^5	10 000
	195	-3×10^5	−30 000

comparison of CD spectra of such peptides in the absence and presence of small unilamellar phospholipid vesicles shows a clear difference in the type of secondary structure. Measurements with lipid vesicles are tricky, because their physical extensions give rise to scatter. Other options in this context include CD experiments at lipid monolayers, which can be realised at synchrotron beam lines or by usage of optically clear vesicles (reverse micelles).

Circular dichroism spectroscopy can also be used to monitor changes of secondary structure within a sample over time. Frequently, CD instruments are equipped with temperature control units and the sample can be heated in a controlled fashion. As the protein undergoes its transition from the folded to the unfolded state, the CD at a certain wavelength (usually 222 nm) is monitored and plotted against the temperature, thus yielding a thermal denaturation curve that can be used for stability analysis.

Further applications include the use of CD for kinetic measurements using the stopped flow technique.

2.6 Light scattering

The scattering of light can yield a number of valuable insights into the properties of macromolecules, including the molar mass, gross molecular dimensions and diffusion coefficients, as well as association/dissociation properties and internal dynamics. The incident light hitting a macromolecule is scattered in all directions, with the intensity of the scatter being only about 10^{-5} of the original intensity. The scattered light is measured at angles higher than 0° and less than 180°. Most of the scattered light possesses the same wavelength as the incident light, and this phenomenon is called elastic light scattering. When the scattered light has a wavelength higher or lower than the incident light, the phenomenon is called inelastic light scattering. The special properties of lasers (see Section 1.2), with high monochromaticity, narrow focus and strong intensity, make them ideally suited for light-scattering applications.

2.6.1 Elastic (static) light scattering

Elastic light scattering is also known as Rayleigh scattering and involves measuring the intensity of light scattered by a solution at an angle relative to the incident laser beam. The technique exploits two basic principles:

- The scattering intensity of macromolecules is proportional to the product of molar mass and concentration of the scatterer.
- The angular variation of the scattered light is directly related to the size of the scattering molecule.

As such, it is ideal for determination of the molar mass M of macromolecules, as the contribution of small solvent molecules can be neglected. In an ideal solution, the macromolecules are entirely independent from each other, and the light scattering can be described as the ratio of the scattered and incident light intensity (Rayleigh ratio):

$$\frac{I_\theta}{I_0} \sim R_\theta = P_\theta \times K \times c \times M \tag{2.31}$$

with

$$K = \frac{4\pi^2 n_0^2}{\mathrm{N_A}\lambda^4}\left(\frac{\mathrm{d}n}{\mathrm{d}c}\right)^2, \tag{2.32}$$

where I_θ is the intensity of the scattered light at angle θ, I_0 is the intensity of the incident light, K is a constant proportional to the squared refractive index increment $(\mathrm{d}n/\mathrm{d}c)$ and c is the concentration. R_θ is called the Rayleigh ratio, and P_θ is called the form factor and describes the angular dependence of the scattered light.

For non-ideal solutions, however, interactions between molecules need to be considered. The scattering intensity of real solutions has been calculated by Debye, and takes concentration fluctuations into account. This results in an additional correction term comprising the second virial coefficient B, which is a measure of the strength of interactions between molecules:

$$\frac{K}{R_\theta} = \frac{1}{P_\theta}\left(\frac{1}{M} + 2 \times B \times c\right). \tag{2.33}$$

Determination of molar mass with multi-angle light scattering (MALS)

In solution, there are only three methods for absolute determination of molar mass: membrane osmometry, sedimentation equilibrium centrifugation (see Section 4.1.4) and light scattering. These methods are absolute, because they do not require any reference to molar mass standards. In order to determine the molar mass from light scattering, three parameters must be measured: the intensity of scattered light at each angle, the concentration of the macromolecule and the specific refractive index increment of the solvent. The minimum instrumentation required consists of a light source, a MALS detector and a refractive index detector. These instruments can be used in batch mode but can also be connected to a high-performance liquid chromatography (HPLC) instrument to enable online determination of the molar mass of eluting macromolecules. The chromatography of choice is size-exclusion chromatography (SEC), also called gel filtration, and the combination of these methods is known as SEC-MALS. Unlike conventional SEC, the molar mass determination from MALS is independent of the elution volume of the macromolecule. This is a valuable advantage, as the retention time of a macromolecule on the size-exclusion column can depend on its shape and conformation.

2.6.2 Quasi-elastic (dynamic) light scattering: photon correlation spectroscopy

While the intensity and angular distribution of scattered light yields information about the molar mass and dimension of macromolecules, the wavelength analysis of scattered light allows conclusions as to the transport properties of macromolecules. Due to rotation and translation, macromolecules move into and out of a very small region in the solution. This Brownian motion happens on a timescale of μs to ms, and the translation component of this motion is a direct result of diffusion, which leads to a broader wavelength distribution of the scattered light compared with the incident light. This effect is the subject of dynamic light scattering, and yields the distribution of diffusion coefficients of macromolecules in solution.

The diffusion coefficient is related to the particle size by an equation known as the Stokes–Einstein relationship, which describes the Brownian motion of a particle in solution. The parameter derived is the hydrodynamic radius R_H (also called the Stokes radius), which is the size of a spherical particle that would have the same diffusion coefficient (D) in a solution with the same viscosity (η):

$$R_{\mathrm{H}} = \frac{\mathrm{k_B}T}{6\pi\eta D} \tag{2.34}$$

where $\mathrm{k_B}$ is the Boltzmann constant. Notably, the hydrodynamic radius describes an idealised particle and can differ significantly from the true physical size of a macromolecule. This is certainly true for most proteins that are not strictly spherical, and their hydrodynamic radius thus depends on their shape and conformation.

In contrast to SEC, dynamic light scattering measures the hydrodynamic radius directly and accurately, as the former method relies on comparison with standard molecules and several assumptions.

Applications of dynamic light scattering include determination of diffusion coefficients and assessment of protein aggregation, and can aid many areas in practice. For instance, the development of 'stealth' drugs that can hide from the immune system or certain receptors relies on the 'PEGylation' of molecules. As conjugation with polyethylene glycol (PEG) increases the hydrodynamic size of the drug molecules dramatically, dynamic light scattering can be used as product control and a measure of efficiency of the drug.

A technique related to photon correlation spectroscopy is nanoparticle tracking analysis, which measures the Brownian motion of nanoparticles using the Stokes–Einstein relationship given above. In contrast to photon correlation spectroscopy, the motion of particles in nanoparticle tracking analysis is analysed by video whereby the position of particles is recorded in two dimensions. This mode of analysis is based on the relationship of the product of diffusion coefficient (D) and time (t) with the mean square displacement of a particle ($\overline{<x, y>}^2$):

$$D \times t = \frac{\mathrm{k_B}T}{6\pi\eta R_{\mathrm{H}}} = \frac{\overline{\langle x, y\rangle}^2}{4} \tag{2.35}$$

2.6.3 Inelastic light scattering: Raman spectroscopy

When the incident light beam hits a molecule in its ground state, there is a (low) probability that the molecule is excited and occupies the next higher vibrational state (see Figs 1.3 and 2.36). The energy needed for the excitation is a defined increment, which will be missing from the energy of the scattered light. The wavelength of the scattered light is thus increased by an amount associated with the difference between two vibrational states of the molecule (Stokes lines). Similarly, if the molecule is hit by the incident light in its excited state and transitions to the next lower vibrational state, the scattered light has higher energy than the incident light, which results in a shift to lower wavelengths (anti-Stokes lines). These lines constitute the Raman spectrum (see Section 2.7.3). If the wavelength of the incident light is chosen such that it coincides with an absorption band of an electronic transition in the molecule, there is a significant increase in the intensity of bands in the Raman spectrum. This technique is called resonance Raman spectroscopy.

2.7 Raman and IR spectroscopy

Within the electromagnetic spectrum (see Fig. 1.1), the energy range below the UV/Vis is the IR region, encompassing the wavelength range of about 700 nm to 25 μm, and thus reaching from the red end of the visible to the microwave region. The absorption of IR light by a molecule results in transition to higher levels of vibration (see Fig. 1.3).

For the purpose of this discussion, the bonds between atoms can be considered as flexible springs, illustrating the constant vibrational motion within a molecule (Fig. 2.32). Importantly, it is assumed that the position and orientation of the molecule does not change on the timescale of a vibrational transition. Newtonian mechanics are applied to describe the molecular vibration, and thus amplitudes and speeds of individual atoms in a vibrational mode depend on the spring stiffness, i.e. the bond strength.

Bond vibrations can thus be either stretching or bending (deformation) actions. Theory predicts that a molecule with N atoms will have a total of $3N-6$ fundamental vibrations ($3N-5$, if the molecule is linear): $2N-5$ bending and $N-1$ stretching modes (Fig. 2.33).

Infrared and Raman spectroscopy are often subsumed as rota-vibrational spectroscopy and give similar information about a molecule, but the criteria for the phenomena to occur are different for each type. For non-centrosymmetric molecules, incident IR light will give rise to an absorption band in the IR spectrum, as well as a peak in the Raman spectrum. However, as shown in Fig. 2.33, centro-symmetric molecules, such as for example CO_2, show a selective behaviour: bands that appear in the IR spectrum do not appear in the Raman spectrum, and vice versa.

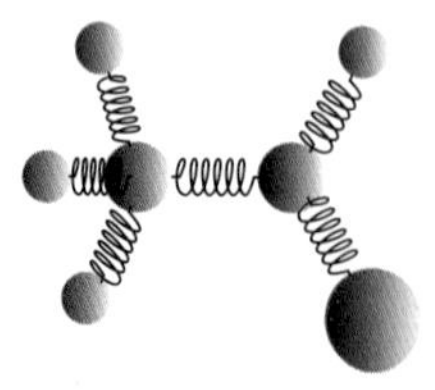

Fig. 2.32. Acetaldehyde vibrational model.

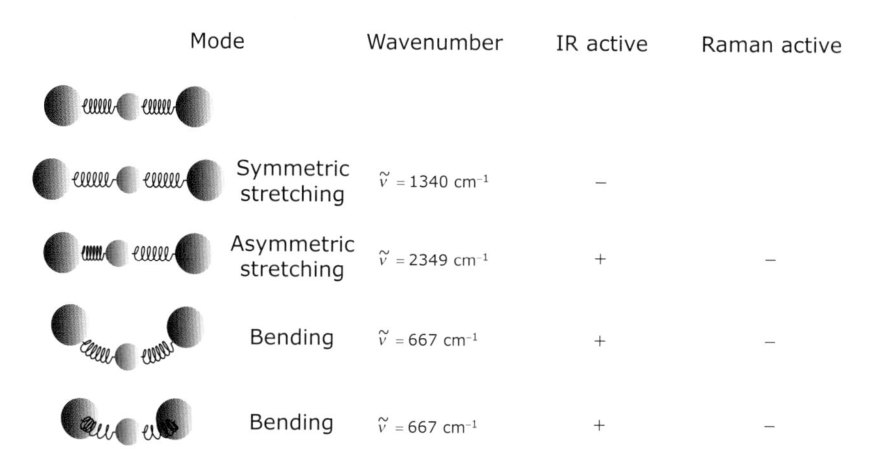

Fig. 2.33. Normal vibration modes for CO_2. For symmetric molecules, bands that are active in the IR are not active in the Raman spectrum.

An IR spectrum arises from the fact that a molecule absorbs incident light of a certain wavelength, which will then be 'missing' from the transmitted light. The recorded spectrum will show an absorption band.

A Raman spectrum arises from the analysis of scattered light, and we have already introduced the basics of inelastic light scattering in Section 2.6. The largest part of an incident light beam passes through the sample (transmission). A small part is scattered isotropically, i.e. uniformly in all directions (Rayleigh scatter), and possesses the same wavelength as the incident beam. The Raman spectrum arises from the fact that a very small proportion of light scattered by the sample will have a different frequency from the incident light. As different vibrational states are excited, energy portions will be missing, thus giving rise to peaks at lower frequencies than the incident light (Stokes lines). Notably, higher frequencies are also observed (anti-Stokes lines); these arise from excited molecules returning to ground state. The emitted energy is dumped onto the incident light which results in scattered light of higher energy than the incident light.

The criterion for a band to appear in the IR spectrum is that the transition to the excited state is accompanied by a change in dipole moment, i.e. a change in charge displacement. Conversely, the criterion for a peak to appear in the Raman spectrum is a change in polarisability of the molecule during the transition.

2.7.1 Vibration and rotation

As mentioned above, the vibrational modes between two bonded atoms can be described using Newtonian mechanics. Using the analogy with a spring, the force (F) required to extend the spring (interatomic bond) is given by Hooke's law:

$$F = -k \times x, \tag{2.36}$$

where k is the force constant describing the stiffness of the spring (bond) and x is the elongation distance.

The elongation as a function of time $x(t)$ is given by the harmonic oscillator:

$$x(t) \sim \cos(2\pi \nu t), \tag{2.37}$$

with the frequency:

$$\nu = \frac{1}{2\pi}\sqrt{\frac{k}{\mu}}. \tag{2.38}$$

The oscillation frequency (equivalent to the frequency of light required to activate this mode) is a function of the spring (bond) stiffness represented by k, and the masses of the two connected objects (atoms), represented by the reduced mass μ:

$$\mu = \frac{m_A \times m_B}{m_A + m_B} \tag{2.39}$$

or

$$\frac{1}{\mu} = \frac{1}{m_A} + \frac{1}{m_B}. \tag{2.40}$$

As the atomic mass is a parameter that determines the frequency of the photon absorbed or scattered for this mode, it follows that isotope exchange within a molecule leads to small changes in IR or Raman spectra.

The potential energy V of the harmonic oscillator describes a quadratic (parabolic) function, which yields higher energies for a system the higher the elongation (bond lengthening). x_0 describes the spring length (bond distance) where the system possesses its lowest energy. The force constant k is the second derivative of the potential (bond) energy:

$$V \sim k \times (x - x_0)^2, \tag{2.41}$$

and thus:

$$k = \frac{d^2V}{dx^2}. \tag{2.42}$$

Whereas a Newtonian harmonic oscillator can take any potential energy V, this is not possible for an atomic system due to the quantisation of

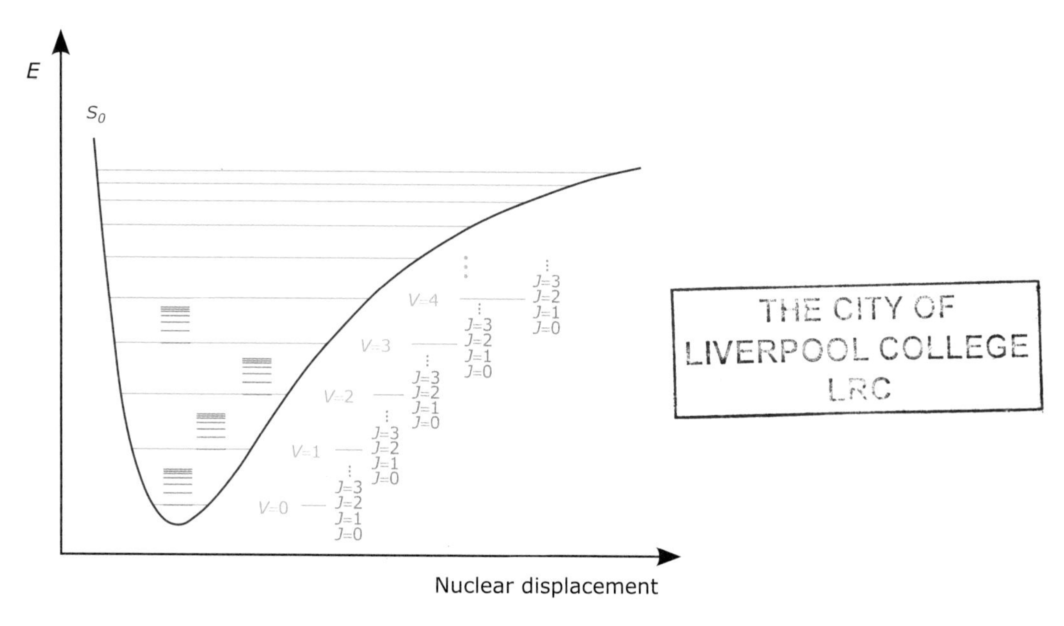

Fig. 2.34. Rotational states (described by the quantum number J; yellow) are overlayed onto each vibrational state (quantum number v; green) of a molecule. The blue curve represents the potential energy of the molecule in the electronic ground state (S_0). Where the curve follows the shape of a parabolic function (i.e. at low rota-vibrational energy levels), the assumption of a harmonic oscillator can be applied. At higher energies, other potential functions such as the Morse potential need to be used (Morse, 1929).

atomic phenomena. The allowed energy levels of a quantum mechanical harmonic oscillator are discretely spaced and can be calculated for a given vibrational quantum number v by:

$$E_{\text{vib}} = \left(v + \frac{1}{2}\right) h\nu \tag{2.43}$$

where h is the Planck constant and ν the frequency of the electromagnetic radiation.

A similar case can be made for the Newtonian mechanical description of a rotating system. For quantum mechanical treatment, the rotational quantum number J is introduced; rotational levels are overlayed onto the vibrational states of a molecule (Fig. 2.34). The discrete energies for a rigid rotor (constant bond length) are given by

$$E_{rot} = J(J+1)B = J(J+1)hc\overline{B} \tag{2.44}$$

Table 2.14. Gross selection rules for rota-vibrational transitions in IR and Raman spectroscopy.

IR selection rule	Raman selection rule
A rota-vibrational transition is IR active if there is:	A rota-vibrational transition is Raman active if there is:
• a change in dipole moment • a change in the quantum numbers $\Delta v = +1$, $\Delta J = \pm 1$	• a change in polarisability • a change in the quantum numbers $\Delta v = +1$, $\Delta J = 0, \pm 2$
Typically observed with asymmetrical modes	Typically observed with symmetrical modes

where B is the rotational constant and $\overline{B}$ is defined as:

$$\overline{B} = \frac{h}{8\pi^2 c \mu r^2}, \tag{2.45}$$

with $\overline{B}$ having the unit of wavenumbers: $[\overline{B}] = 1\ \text{cm}^{-1}$, and where c is the speed of light, μ is the reduced mass and r is the bond distance between the two atoms of the rotor.

In a rigid rotor, the bond length does not change and the energy difference between two rotational levels is thus constant and equal to $2B$. An added complication arises from the fact that rotating molecules also vibrate and thus have variable bond distances; they are not rigid rotors. This means that the rotational constant B depends on the vibrational state; the higher the vibrational state, the lower the value of B (higher vibrational states have larger bond distances r). For the non-rigid rotor, the rotational constant thus needs to be corrected for varying bond distances depending on the vibrational levels (vibrational–rotational interaction constant). In the rotational spectrum (see Fig. 2.36), the peaks of the individual transitions are thus not spaced equally but show slightly decreasing distances with increasing rotational quantum numbers J.

The quantum mechanical rules will also need to be considered when assessing whether a rota-vibrational transition appears in either IR or Raman spectroscopy. The gross selection rules are therefore as shown in Table 2.14.

The rule of mutual exclusion states that, for a centro-symmetric molecule (i.e. a molecule that possesses an inversion centre), a vibrational mode may be observed either in IR or in Raman spectroscopy, but not in both.

2.7.2 IR spectroscopy

The fundamental frequencies observed are characteristic of the functional groups concerned, hence the term fingerprint. Figure 2.35 shows the major bands of a Fourier transform IR (FTIR) spectrum of starch. As the number

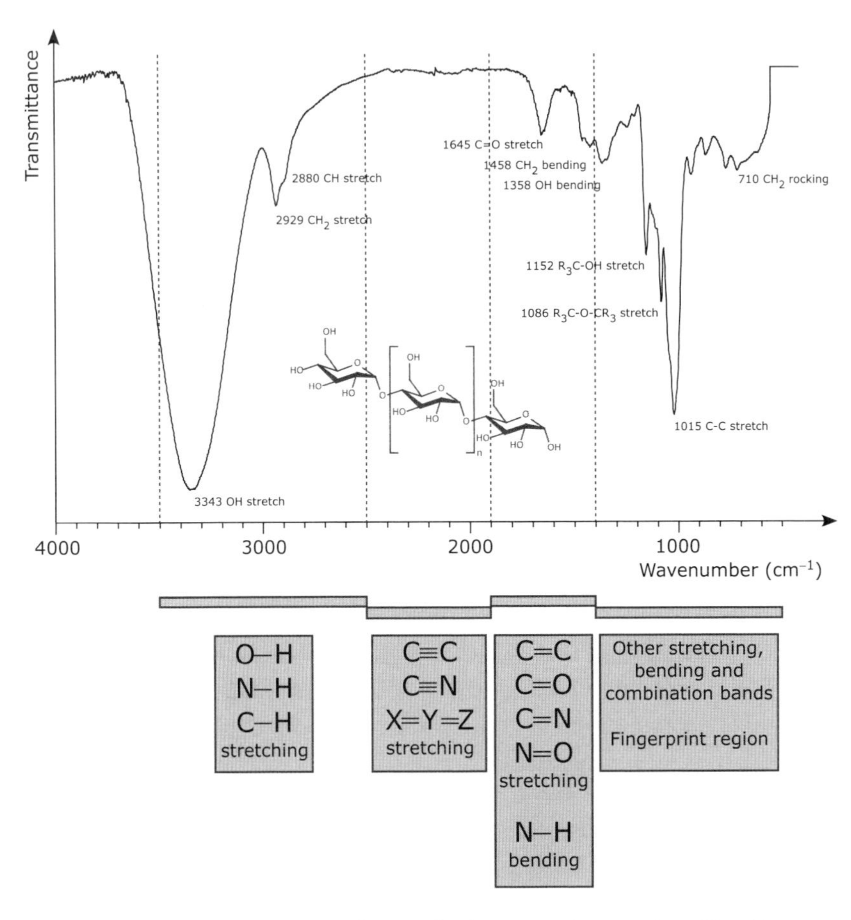

Fig. 2.35. IR absorption regions illustrated using a Fourier transform IR with attenuated total reflection (FTIR-ATR) spectrum of starch, a carbohydrate consisting of glucose units linked by glycosidic bonds. Plant starch is a mixture of the unbranched polysaccharide amylose (~25%; structure shown in figure) and the branched amylopectin (~75%); animal starch contains even more branched glycogen instead of amylopectin. Note the C=O stretching mode at 1645 cm^{-1}, which indicates the existence of the open aldose form of glucose at the ends of individual macromolecules. (Spectrum courtesy of Lawren Sullivan, Griffith University).

of functional groups increases in more complex molecules, the absorption bands become more difficult to assign. However, groups of certain bands regularly appear near the same wavelength and may be assigned to specific functional groups. Such group frequencies are thus extremely helpful in structural diagnosis. A more detailed analysis of the structure of a molecule is possible, because the wavenumber associated with a particular functional group varies slightly, owing to the influence of the molecular environment. For example, it is possible to distinguish between C–H vibrations in methylene ($-CH_2$) and methyl groups ($-CH_3$).

2.7.3 Raman spectroscopy

Whereas IR spectroscopy is an absorption spectroscopy that requires a vibrational mode of the molecule to cause a change in the dipole moment, the Raman effect is due to a momentary distortion of electrons around a molecular bond upon incident light. The induced dipole and temporary polarisation disappears upon emission of light; Raman spectroscopy is thus a scattering spectroscopy (see also Section 2.6.3).

When a molecule is hit by incident light, the strongest scattering effect is the Rayleigh scatter (Section 2.6.3); this is elastic scattering, i.e. the energy of the scattered photon is the same as the energy of the absorbed photon. Raman scattering is inelastic: the scattered photons have different energies to the incident photon. The incident light excites the molecule into the first electronic excited state S_1 (Fig. 2.36). As the molecule returns to the electronic ground state S_0, it may end up in an excited rotational state of the electronic ground state (e.g. excitation: $S_0v_0J_0 \rightarrow S_1v_1J_0$; emission: $S_1v_1J_0 \rightarrow S_0v_0J_2$). This transition results in re-emission of a photon that has less energy than the incident photon and is called a Stokes line. Anti-Stokes lines are of less intensity than Stokes lines and appear if the re-emitted photon has higher energy than the incident photon (e.g. excitation: $S_0v_0J_1 \rightarrow S_1v_1J_3$; emission: $S_1v_1J_2 \rightarrow S_0v_0J_0$). As the molecular energy

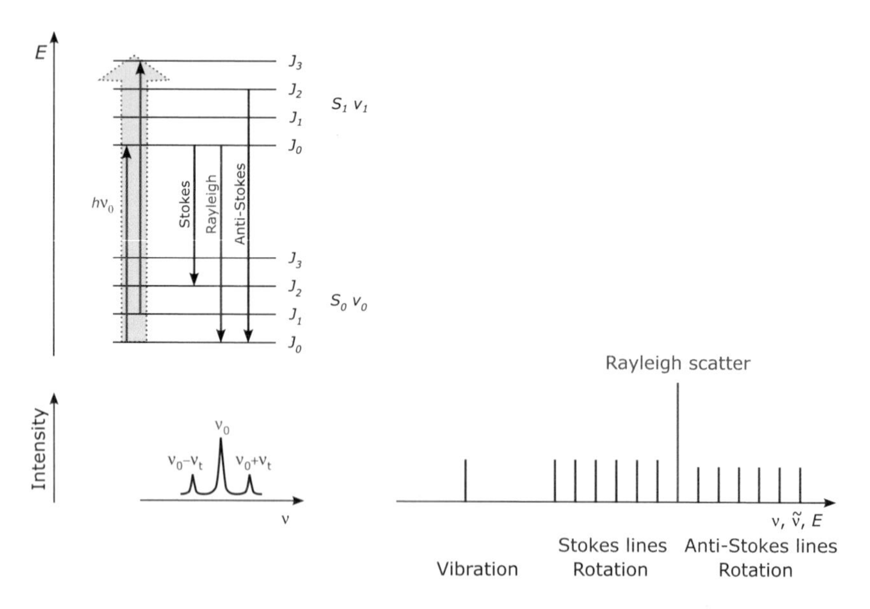

Fig. 2.36. Molecular transitions in Raman scattering (left top) that lead to the appearance of Raman peaks (left bottom). A schematic rota-vibrational Raman spectrum is shown on the bottom right.

levels are quantised, Raman spectroscopy produces discrete lines that image the energetic rota-vibrational structure of a molecule.

From Figure 2.36, it is clear that the information of the Raman spectrum is in the shift of the frequency (or wavenumber) of an observed line with respect to the frequency of the incident light (ν_0). Therefore, Raman wavenumbers of particular modes are obtained as differences with respect to the incident light. Importantly, the magnitude of the shift is independent of the energy of the incident light. The same Raman spectrum is thus obtained with lasers of different wavelengths. The intensity of the scattered light in the Raman experiment is in the order of 0.001% of the intensity of the incident light and can be calculated from the following equation:

$$I = \frac{8\pi^4 \alpha^2}{\lambda^4 r^2}(1 + \cos^2\theta)I_0. \quad (2.46)$$

Here, α is the polarisability of the molecule, λ is the wavelength, I_0 is the intensity of incident radiation, θ is the angle between the incident and scattered light beam and r is the distance to detector.

In order to observe the rather weak Raman scattering, powerful laser sources are required. With some molecules, fluorescence is observed as an unwanted side effect. Other possible disadvantages could be heating and/or photodecomposition of the sample. However, as water is a weak Raman scatterer, samples in aqueous solution can be measured, which is a distinct advantage over IR spectroscopy. Similarly, whereas IR as an absorption spectroscopy requires translucent samples, non-translucent samples can be subjected to Raman spectroscopy, because in the latter the scattered light is analysed. With lasers being the light source for Raman spectroscopy, it is further possible to achieve small beam diameters and thus either analyse small sample areas or even conduct lateral surface mapping.

The analysis of Raman spectra includes consideration of peak position, intensity and form, as well as polarisation (Table 2.15). This allows identification of the type of symmetry of individual vibrations, but not the

Table 2.15. Parameters analysed in Raman spectroscopy.

Parameter	Information content
Wavenumber[a] of Raman peak	Composition of material
Change of wavenumber[a] of Raman peak	Stress or strain
Polarisation	Crystal symmetry, orientation
Width of peak	Quality of crystal
Intensity	Amount of material

[a] Note that this is actually a wavenumber difference with respect to the incident light.

determination of structural elements of a molecule. The depolarisation is calculated as the ratio of two intensities with perpendicular and parallel polarisation with respect to the incident beam. The use of lasers as a light source for Raman spectroscopy easily facilitates the use of linearly polarised light. Practically, the Raman spectrum is measured twice. In the second measurement, the polarisation plane of the incident beam is rotated by 90°.

2.7.4 Instrumentation

Dispersive IR spectrometers

The most common source for IR light is white-glowing zircon oxide or the so-called globar made of silicon carbide with a glowing temperature of 1500 K. The beam of IR light passes a monochromator and splits into two separate beams: one runs through the sample, the other through a reference made of the substance the sample is prepared in. After passing through a splitter alternating between both beams, they are reflected into the detector. The reference is used to compensate for fluctuations in the source, as well as to cancel possible effects of the solvent. Samples of solids are either prepared as a thick suspension (mull) in oil such as nujol and held as a layer between NaCl plates, or mixed with solid KBr and pressed into discs. Non-covalent materials must be used for sample containment and in the optics, as these materials are transparent to IR. All materials need to be free of water, because of the strong absorption of the O–H vibration.

Fourier transform IR (FTIR) spectrometers

Analysis using a Michelson interferometer enables FTIR spectroscopy. The polychromatic light emitted from the source is passed into the interferometer (Fig. 2.37) and then split into two beams that are reflected back onto the point of split (interferometer plate). Using a movable mirror, path length differences are generated between both beams yielding an interferogram that has all wavelengths 'encoded'. The beam then enters the sample compartment where it is transmitted through or reflected off the surface of the sample, depending on the type of analysis being performed. This is where specific wavelengths, which are characteristic of the sample, are absorbed. The interferogram is recorded by the detector and can be deconvoluted into a conventional IR spectrum by Fourier transformation. As the interferometer signal provides the entire wavelength range to be scanned simultaneously, the IR spectrum can be measured very quickly, usually in the order of seconds, thus providing a clear advantage over the conventional scanning method in dispersive instruments. A huge number of repeats taken can thus be accumulated and hence improve the quality of the spectrum.

FTIR spectrometers with attenuated total reflection

Attenuated total reflection (ATR) employs the features of an evanescent field, the same technology used in surface plasmon resonance (see Section 5.1.1). The IR light enters the internal reflection element (a glass prism similar to the one described for surface plasmon resonance) and probes the sample on the surface of the prism (Fig. 2.37, inset). The fact that the sample is probed multiple times has the advantage of yielding stronger absorbances. However, due to scatter occurring at each reflection, quantification of the observed absorbances is difficult. FTIR-ATR has become a very popular method, as small amounts of solid and liquid samples can be measured conveniently without lengthy preparation.

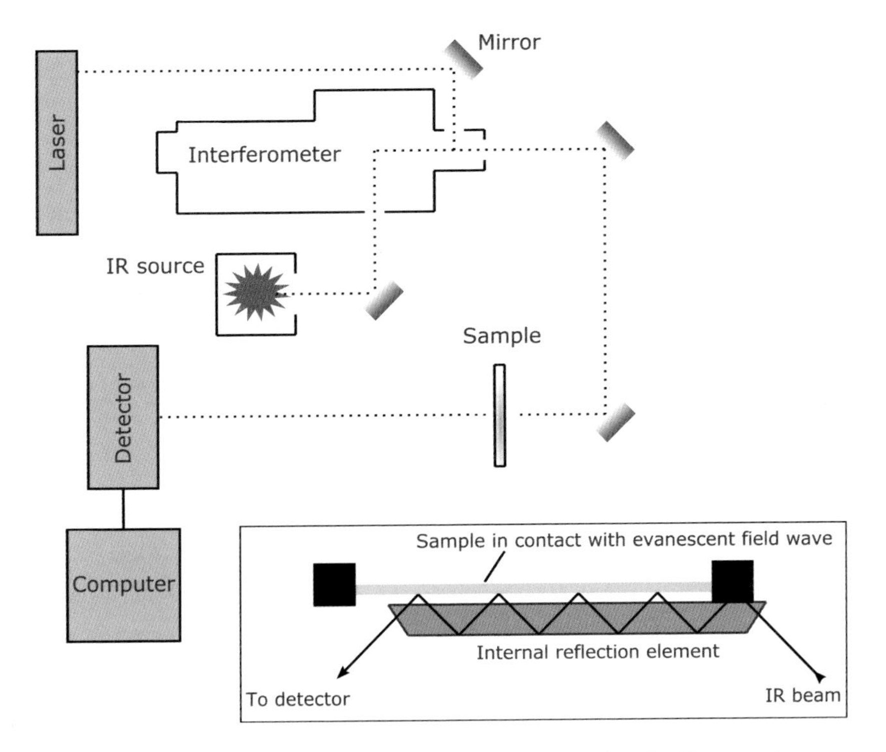

Fig. 2.37. Schematic layout of an FTIR instrument. Note that the IR source is polychromatic and the mirror movement in the interferometer yields the light wavelength range. Pulsed laser light is required to provide a time base for the interferometer. Inset: Schematics of attenuated total reflection (ATR) in FTIR instruments.

Raman spectrometers

The Raman effect can in principle be observed with bright, monochromatic light of any wavelength; however, light from the visible region of the electromagnetic spectrum is normally used due to few unwanted absorption effects, but UV and NIR wavelengths can also be employed. As outlined above, the ideal light source for Raman spectrometers are lasers. UV lasers (244 and 325 nm) are typically used for biological samples, and visible lasers (Ar^+ laser: 488 and 514 nm; Kr^+ laser: 531 and 617 nm; He-Ne laser: 633 nm) are the most frequently used light sources for general purpose analyses. NIR lasers (diode lasers: 785 and 830 nm; Nd-YAG laser: 1024 nm) are in use for some polymer and biological samples. Shorter wavelengths have the clear advantage of increased Raman scattering efficiency, as the intensity of Raman bands varies with the fourth power of the frequency of the incident light. However, the risk of fluorescence background increases, and sample damage through heating needs to be considered.

For Raman spectroscopy, samples can be in the form of gases (in sealed glass tubes), liquids (in glass vials, aqueous solutions possible) or solids (pellets, powders). Because the Raman effect is observed in light scattered off the sample, typical spectrometers use a 90° configuration (Fig. 2.38). In order to separate the weak Raman lines from the intense Rayleigh scatter, high-quality wavelength selection devices need to be employed. Dispersive instruments use double or triple grating monochromators, often combined with holographic interference filters (notch filters) or holographic gratings. Fourier transform-Raman spectrometers are also in use.

Raman spectrometers can also be combined with microscopes, thereby achieving a high spatial resolution that enables lateral mapping.

2.7.5 Applications

The use of IR and Raman spectroscopy is mainly in chemical and biochemical research of small compounds such as drugs, metabolic intermediates and substrates. Examples are the determination of bond lengths from rota-vibrational analysis, identification of pure compounds and identification of sample constituents (e.g. in food) when coupled to a separating method such as gas chromatography. Both spectroscopies are thus routinely applied in chemical and mineralogical laboratories, as well as in counterfeit and forensic applications. Hand-held spectrometers also allow mobile applications such as threat screening. Most routine applications are based on empirical comparison of spectra obtained from a sample with a database of spectra from pure components.

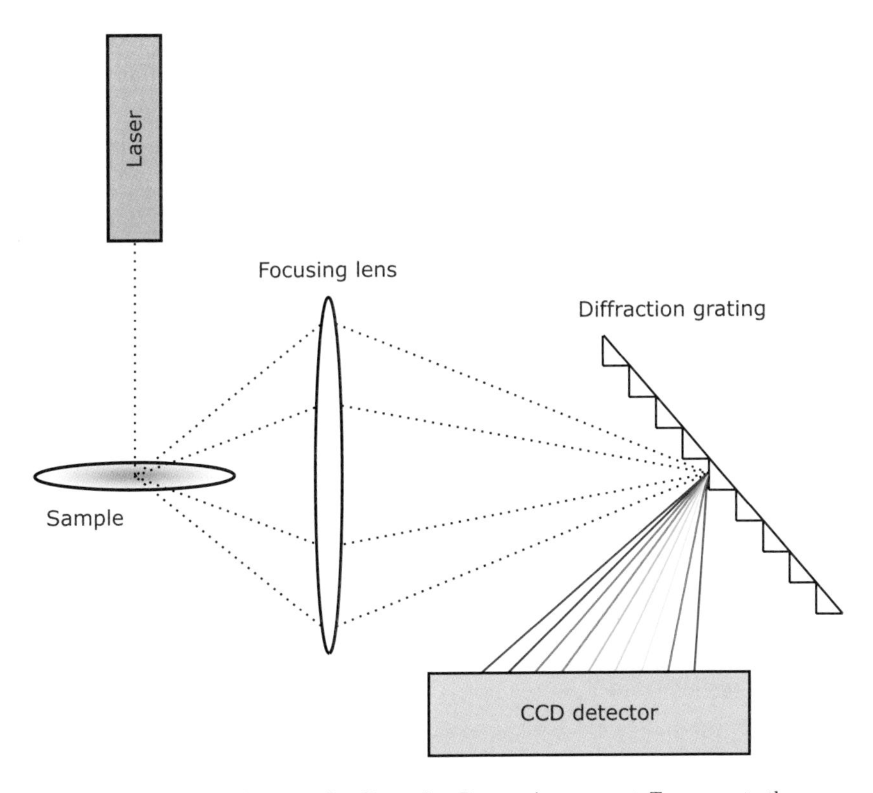

Fig. 2.38. Schematic layout of a dispersive Raman instrument. To separate the collected Raman scattered light into individual wavelengths, it is focused onto a grating that separates the light into different wavelengths (diffraction). The dispersed beam is then directed to the charge-coupled device (CCD) detector to be acquired.

There is increasing use of FTIR for analysis of peptides and proteins. The peptide bond gives rise to nine characteristic bands, named amide A, B, I, II and III–VII. The amide I (1600–1700 cm^{-1}) and amide II (1500–1600 cm^{-1}) bands are the major contributors to the protein IR spectrum. Both bands are directly related to the backbone conformation and have thus been used for assessment of the secondary structure of peptides and proteins. The interpretation of spectra of molecules with a large number of atoms usually involves deconvolution of individual bands and second-derivative spectra.

Time-resolved FTIR enables the observation of protein reactions at the submillisecond timescale. This technique has been established by investigation of the light-driven proton pump bacteriorhodopsin. The catalytic steps in the proton-pumping mechanism of bacteriorhodopsin that were successfully investigated with this technique involve transfer of a proton from the Schiff base ($R_1R_2C{=}N\text{-}R_3$) to a catalytic aspartate residue, followed by reprotonation of a second catalytic aspartate residue.

FURTHER READING

Atomic spectroscopy

Bings, N. H., Bogaerts, A. & Broekaert, J. A. C. (2010). Atomic spectroscopy: a review. *Analytical Chemistry* 82, 4653–81.

Butler, O. T., Cairns, W. R. L., Cook, J. M. & Davidson, C. M. (2013). Atomic spectrometry update. Environmental analysis. *Journal of Analytical Atomic Spectrometry* 28, 177–216.

Kaulich, B., Gianoncelli, A., Beran, A. *et al.* (2009). Low-energy X-ray fluorescence microscopy opening new opportunities for bio-related research. *Journal of the Royal Society Interface* 6 **Suppl. 5**, S641–7.

Websites on atomic spectroscopy

Photon Emission Applet: http://zebu.uoregon.edu/nsf/emit.html

X-ray Fluorescence Microscopy for Biology and Bionanotechnology: Challenges and Unique Opportunities: http://erl.chess.cornell.edu/gatherings/2011_Workshops/talks/WS5Vogt.pdf

UV/Vis spectroscopy

Pretsch, E., Bühlmann, P. & Badertscher, M. (2009). *Structure Determination of Organic Compounds*, 4th edn. Berlin, Heidelberg: Springer Verlag. (A compendium with tables of spectral data.)

Schoonheydt, R. A. (2010). UV-VIS-NIR spectroscopy and microscopy of heterogeneous catalysts. *Chemical Society Reviews* 39, 5051–66.

Simonian, M. H. & Smith, J. A. (2006). Spectrophotometric and colorimetric determination of protein concentration. In *Current Protocols in Molecular Biology*. New York: Wiley Interscience.

Websites on UV/Vis spectroscopy

UV-Vis Absorption Spectroscopy: Theoretical principles: http://teaching.shu.ac.uk/hwb/chemistry/tutorials/molspec/uvvisab1.htm

Visible and Ultraviolet Spectroscopy: http://www.cem.msu.edu/~reusch/VirtualText/Spectrpy/UV-Vis/spectrum.htm#uv1

Visual Quantum Mechanics: Absorption Spectroscopy: http://phys.educ.ksu.edu/vqm/html/absorption.html

Fluorescence spectroscopy

Beausang, J. F., Sun, Y., Quinlan, M. E., Forkey, J. N. & Goldman, Y. E. (2012). Orientation and rotational motions of single molecules by polarised total internal reflection fluorescence microscopy (polTIRFM). *Cold Spring Harbor Protocols* 2012, 535–45.

Brown, M. P. & Royer, C. (1997). Fluorescence spectroscopy as a tool to investigate protein interactions. *Current Opinion in Biotechnology* **8**, 45–9.

Groemping, Y. & Hellmann, N. (2005). Spectroscopic methods for the determination of protein interactions. *Current Protocols in Protein Science*, Chapter 20, Unit 20.8.

Hwang, L. C. & Wohland, T. (2007). Recent advances in fluorescence cross-correlation spectroscopy. *Cell Biochemistry and Biophysics* **49**, 1–13.

Jameson, D. M. & Ross, J. A. (2010). Fluorescence polarisation /anisotropy in diagnostics and imaging. *Chemical Reviews* **110**, 2685–708.

Lakowicz, J. R. (1999). *Principles of Fluorescence Spectroscopy*, 2nd edn. New York: Kluwer Academic/Plenum Publishers. (An authorative text book on fluorescence spectroscopy.)

Langowski, J. (2008). Protein–protein interactions determined by fluorescence correlation spectroscopy. *Methods in Cell Biology* **85**, 471–84.

Prinz, A., Reither, G., Diskar, M. & Schultz, C. (2008). Fluorescence and bioluminescence procedures for functional proteomics. *Proteomics* **8**, 1179–96.

Roy, R., Hohng, S. & Ha, T. (2008). A practical guide to single-molecule FRET. *Nature Methods* **5**, 507–16.

Van Engelenburg, S. B. & Palmer, A. E. (2008). Fluorescent biosensors of protein function. *Current Opinion in Chemical Biology* **12**, 60–5.

Winkler, J. R. (2013). FRETting over the spectroscopic ruler. *Science* **339**, 1530–1. (A commentary on recent experimental data that electron transfer can compete with FRET.)

Websites on fluorescence spectroscopy

Fluorescence Tutorials: http://www.invitrogen.com/site/us/en/home/support/Tutorials.html

Fluorescence Microscopy – Excitation Balancer for multiply labelled specimen: http://www.microscopyu.com/tutorials/java/fluorescence/excitationbalancer/index.html

Luminometry

Bacart, J., Corbel, C., Jockers, R., Bach, S. & Couturier, C. (2008). The BRET technology and its application to screening assays. *Journal of Biotechnology* **3**, 311–24.

Deshpande, S.S. (2001). Principles and applications of luminescence spectroscopy. *Critical Reviews in Food Science and Nutrition* **41**, 155–224.

Jia, Y., Quinn, C. M., Kwak, S. & Talanian, R. V. (2008). Current in vitro kinase assay technologies: the quest for a universal format. *Current Drug Discovery Technologies* 5, 59–69.

Meaney, M. S. & McGuffin, V. L. (2008). Luminescence-based methods for sensing and detection of explosives. *Analytical and Bioanalytical Chemistry* 391, 2557–76.

Website on luminometry

An Introduction to Chemiluminescence and Bioluminescence Measurements: http://www.comm-tec.com/Library/Tutorials/CTD/Chemiluminescence%20and%20Bioluminescence%20Measurements%20.pdf

Circular dichroism spectroscopy

Barron, L. (2004). *Molecular Light Scattering and Optical Activity*, 2nd edn. Cambridge: Cambridge University Press.

Fasman, G. D. (1996). *Circular Dichroism and the Conformational Analysis of Biomolecules*, 1st edn. New York: Plenum Press. (An authorative text book on circular dichroism in biochemistry.)

Gottarelli, G., Lena, S., Masiero, S., Pieraccini, S. & Spada, G. P. (2008). The use of circular dichroism spectroscopy for studying the chiral molecular self-assembly: an overview. *Chirality* 20, 471–485.

Greenfield, N. J. (2006). Using circular dichroism spectra to estimate protein secondary structure. *Nature Protocols* 1, 2876–90.

Kelly, S. M. & Price, N. C. (2006). Circular dichroism to study protein interactions. *Current Protocols in Protein Science*, Chapter 20, Unit 20.10.

Lyle, R. E. & Lyle, G. G. (1964). A brief history of polarimetry. *Journal of Chemical Education* 41, 308–13.

Martin, S. R. & Schilstra, M.J. (2008). Circular dichroism and its application to the study of biomolecules. *Methods in Cell Biology* 84, 263–93.

Pessoa, J. C., Correia, I., Goncalves, G. & Tomaz, I. (2009). Circular dichroism in coordination compounds. *Journal of the Argentine Chemical Society* 97, 151–65.

Snatzke, G. (1968). Circular dichroism and optical rotatory dispersion – principles and application to the investigation of the stereochemistry of natural products. *Angewandte Chemie International Edition* 7, 14–25.

Snatzke, G. (1979). Circular dichorism and absolute conformation: application of qualitative MO theory to chiroptical phenomena. *Angewandte Chemie International Edition* 18, 363–77.

Websites on CD spectroscopy

DICHROWEB online CD analysis: http://dichroweb.cryst.bbk.ac.uk/
Circular dichroism: http://www.ap-lab.com/circular_dichroism.htm
ChemWiki: Chirality: http://chemwiki.ucdavis.edu/Organic_Chemistry/Chirality

Light scattering

Lindner, P. & Zemb, T. (2002). *Neutron, X-rays and Light. Scattering Methods Applied to Soft Condensed Matter.* Revised subedn. North-Holland Delta Series. The Netherlands: Elsevier. (In-depth coverage of theory and applications of light scattering at expert level.)

Villari, V. & Micali, N. (2008). Light scattering as spectroscopic tool for the study of disperse systems useful in pharmaceutical sciences. *Journal of Pharmaceutical Sciences* **97**, 1703–30.

Websites on light scattering

Laser light scattering: http://www.ap-lab.com/light_scattering.htm
Dynamic light scattering: http://mxp.physics.umn.edu/s05/projects/s05lightscattering/default.htm

Infrared spectroscopy

Beekes, M., Lasch, P. & Naumann, D. (2007). Analytical applications of Fourier transform-infrared (FT-IR) spectroscopy in microbiology and prion research. *Veterinary Microbiology* **123**, 305–19.

Ganim, Z., Chung, H. S., Smith, A. W., *et al.* (2008). Amide I two-dimensional infrared spectroscopy of proteins. *Accounts of Chemical Research* **41**, 432–41.

Hind, A. R., Bhargava, S. K. & McKinnon, A. (2001). At the solid/liquid interface: FTIR/ATR – the tool of choice. *Advances in Colloid and Interface Science* **93**, 91–114.

Kazarian, S. G. & Chan, K. L. A. (2006). Applications of ATR-FTIR spectroscopic imaging to biomedical samples. *Biochimica et Biophysica Acta – Biomembranes* **1758**, 858–67.

Pretsch, E., Bühlmann, P. & Badertscher, M. (2009). *Structure Determination of Organic Compounds*, 4th edn. Berlin, Heidelberg: Springer Verlag. (A compendium with tables of spectral data.)

Tonouchi, M. (2007). Cutting-edge terahertz technology. *Nature Photonics* **1**, 97–105.

Websites on IR spectroscopy

Infrared spectroscopy: a tutorial: http://www.cem.msu.edu/~reusch/VirtualText/Spectrpy/InfraRed/infrared.htm

Organic Chemistry at CU Boulder: IR spectroscopy tutorial: http://orgchem.colorado.edu/hndbksupport/irtutor/tutorial.html

Raman spectroscopy

Benevides, J. M., Overman, S. A. & Thomas, G. J. Jr. (2004). Raman spectroscopy of proteins. In *Current Protocols in Protein Science*, Chapter 17, Unit 17.8.

Wen, Z. Q. (2007). Raman spectroscopy of protein pharmaceuticals. *Journal of Pharmaceutical Sciences* **96**, 2861–78.

Website on Raman spectroscopy

Raman spectroscopy at the University of Bath: http://people.bath.ac.uk/pysdw/newpage11.htm

3 Structural methods

3.1 Electron paramagnetic resonance

Prior to any detailed discussion of electron paramagnetic resonance (EPR) and nuclear magnetic resonance (NMR) methods, it is worthwhile considering the more general phenomena applicable to both.

3.1.1 Magnetic phenomena

Magnetism arises from the motion of charged particles, and, for the purpose of this discussion, the major contribution to magnetism in molecules is due to the spin of the charged particle.

In chemical bonds of a molecule, the negatively charged electrons have a spin controlled by strict quantum rules. A bond is constituted by two electrons with opposite spins occupying the appropriate molecular orbital. According to the Pauli principle, the two electrons must have opposite spins, leading to the term 'paired electrons'. Each of the spinning electronic charges generates a magnetic effect, but in electron pairs this effect is almost self-cancelling. In atoms, a value for magnetic susceptibility may be calculated and is of the order of $-10^{-6}\,g^{-1}$. This diamagnetism is a property of all substances, because they all contain the miniscule magnets, i.e. electrons. Diamagnetism is temperature independent.

If an electron is unpaired, there is no counterbalancing opposing spin and the magnetic susceptibility is of the order of $+10^{-3}$ to $+10^{-4}\,g^{-1}$. The effect of an unpaired electron exceeds the 'background' diamagnetism and gives rise to paramagnetism. The most notable example is certainly the paramagnetism of metals such as iron, cobalt and nickel, which are the materials that permanent magnets are made of. The paramagnetism of these metals is called ferromagnetism. In biochemical investigations, systems with free electrons (radicals) are frequently used as probes.

Similar arguments can be made regarding atomic nuclei. The nucleus of an atom is composed of protons and neutrons, and has a net charge that is normally compensated by the extranuclear electrons. The number of all nucleons (Z) is the sum of the number of protons (P) and the number of neutrons (N). P and Z determine whether a nucleus will exhibit paramagnetism. Carbon-12 (^{12}C), for example, consists of six protons ($P = 6$) and six neutrons ($N = 6$) and thus has $Z = 12$. P and Z are even, and therefore the ^{12}C nucleus possesses no nuclear magnetism. Another example of a nucleus with no residual magnetism is oxygen-16 (^{16}O). All other nuclei with P and Z being uneven possess residual nuclear magnetism.

The way in which a substance behaves in an externally applied magnetic field allows a distinction between dia- and paramagnetism. A paramagnetic material is attracted by an external magnetic field, while a diamagnetic substance is rejected. This principle is employed by the Guoy balance, which allows quantification of magnetic effects. A balance pan is suspended between the poles of a suitable electromagnet supplying the external field. The substance under test is weighed in air with the current switched off. The same sample is then weighed again with the current (i.e. external magnetic field) turned on. A paramagnetic substance appears to weigh more, while a diamagnetic substance appears to weigh less.

3.1.2 The resonance condition

In both EPR and NMR techniques, two possible energy states exist for either electronic or nuclear magnetism in the presence of an external magnetic field. In the low-energy state, the field generated by the spinning charged particle is parallel to the external field. Conversely, in the high-energy state, the field generated by the spinning charged particle is anti-parallel to the external field. When enough energy is input into the system to cause a transition from the low- to the high-energy state, the condition of resonance is satisfied. Energy must be absorbed as a discrete dose (quantum) $\mathrm{h}\nu$, where h is the Planck constant and ν is the frequency. The quantum energy required to fulfil the resonance condition and thus enable transition between the low- and high-energy states may be quantified as:

$$\mathrm{h}\nu = g\beta B, \quad (3.1)$$

where g is a constant called the spectroscopic splitting factor, β is the magnetic moment of the electron (termed the Bohr magneton) and B is the strength of the applied external magnetic field. The frequency ν of the absorbed radiation is a function of the paramagnetic species β and the applied magnetic field B. Thus, either ν or B may be varied to the same effect.

With appropriate external magnetic fields, the frequency of applied radiation for EPR is in the microwave region, and for NMR is in the region of radio frequencies. In both techniques, two possibilities exist for determining the absorption of electromagnetic energy (i.e. enabling the resonance phenomenon):

- A constant frequency ν is applied and the external magnetic field B is swept.
- A constant external magnetic field B is applied and the appropriate frequency ν is selected by sweeping through the spectrum.

For technical reasons, the more commonly used option is a sweep of the external magnetic field.

3.1.3 Principles

The absorption of energy is recorded in the EPR spectrum as a function of the magnetic induction measured in Tesla (T), which is proportional to the magnetic field strength applied. The area under the absorption peak is proportional to the number of unpaired electron spins. Most commonly, the first derivative of the absorption peak is the signal that is actually recorded.

For a delocalised electron, as observed for example in free radicals, the spectroscopic splitting factor g is 2.0023, but for localised electrons such as in transition metal atoms, g varies, and its precise value contains information about the nature of bonding in the environment of the unpaired electron within the molecule. When resonance occurs, the absorption peak is broadened owing to interactions of the unpaired electron with the rest of the molecule ('spin–lattice interactions'). This allows further conclusions as to the molecular structure.

High-resolution EPR may be performed by examining the hyperfine splitting of the absorption peak, which is caused by interaction of the unpaired electron with adjacent nuclei, thus yielding information about the spatial location of atoms in the molecule (Fig. 3.1). The proton hyperfine splitting for free radicals occurs in the range of $0–3\times10^{-3}$ T and yields data analogous to those obtained in high-resolution NMR (see Section 3.2).

The effective resolution of an EPR spectrum can be considerably improved by combining the method with NMR, a technique called electron nuclear double resonance (ENDOR). Here, the sample is irradiated simultaneously with microwaves for EPR and radiofrequencies (RFs) for NMR. The RF signal is swept for fixed points in the EPR spectrum, yielding the EPR signal height versus nuclear RF. This approach is particularly useful when there are a large number of nuclear levels that broaden the normal electron resonance lines.

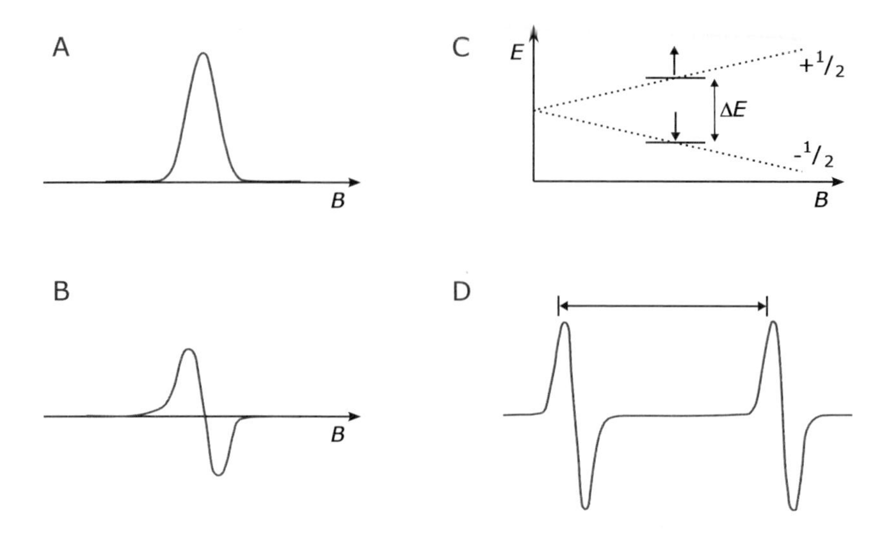

Fig. 3.1. Instead of the absorption signal (A), EPR spectrometry records its first derivative (B). (C) The energy of the two spin states of a free electron are shown as a function of the external magnetic field *B*. Resonance happens when the energy of the applied microwave radiation is the same as the energy difference ΔE. (D) Hyperfine splitting due to coupling of an unpaired electron with a nuclear spin of ½. For the hydrogen atom, the distance between the two signals is 50.7 mT.

The technique of electron double resonance (ELDOR) finds an application in the separation of overlapping multi-radical spectra and to study relaxation phenomena, for example chemical spin exchange. In ELDOR, the sample is irradiated with two microwave frequencies simultaneously. One is used for observation of the EPR signal at a fixed point in the spectrum, while the other is used to sweep other parts of the spectrum. The recorded spectrum is plotted as a function of the EPR signal as a function of the difference of the two microwave frequencies.

3.1.4 Instrumentation

Figure 3.2 shows a diagram of the main components of an EPR instrument. The magnetic fields generated by the electromagnets are of the order of 50–500 mT, and variations of less than 10^{-6} are required for highest accuracy. The monochromatic microwave radiation is produced in a klystron oscillator with wavelengths around 3 cm (corresponding to a frequency of 9 GHz).

The samples are required to be in the solid state; hence, biological samples are usually frozen in liquid nitrogen. The technique is also ideal for investigation of membranes and membrane proteins. Instead of plotting the absorption *A* versus the magnetic field strength *B*, it is the

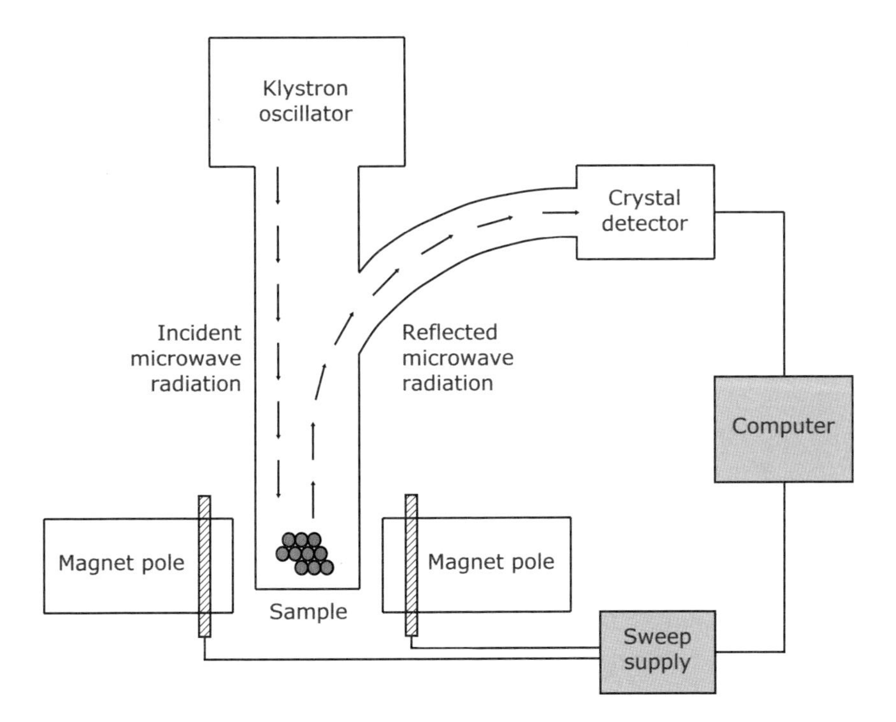

Fig. 3.2. Schematic diagram of an EPR spectrometer.

first-order differential (δA/δ*B*) that is usually plotted against *B* (Fig. 3.1); the resulting shapes are called 'lines' in EPR spectroscopy. Generally, there are relatively few unpaired electrons in a molecule, resulting in fewer than ten lines, which are not closely spaced.

3.1.5 Applications

Metalloproteins

EPR spectroscopy is one of the main methods used to study metalloproteins, particularly those containing molybdenum (xanthine oxidase), copper (cytochrome oxidase, copper blue enzymes) and iron (cytochrome, ferredoxin). Both copper and non-haem iron possess EPR absorption peaks in one of their oxidised states. The appearance and disappearance of their EPR signals are used to monitor the activity of these proteins in the multi-enzyme systems of intact mitochondria and chloroplasts, as well as in isolated enzymes. In many metalloproteins, the ligands coordinating the metal ion are the amino acid residues of the protein. Coordination chemistry requires a specific stereochemical structure of the ligands, and EPR studies show that the geometry is frequently distorted in proteins when compared with model systems. Such distortions may be related to biological function.

Spin labels

Spin labels are stable and non-reactive unpaired electrons used as reporter groups or probes for EPR. The procedure of spin labelling is the attachment of these probes to biological molecules that lack unpaired electrons. The label can be attached to either a substrate or a ligand. Often, a spin label contains the nitric oxide (NO) moiety. These labels enable the study of events that occur with a frequency of 10^7–$10^{11}\,s^{-1}$. If the motion is restricted in some directions, only anisotropic motion (movement in one particular direction) may be studied, for example in membrane-rigid spin labels in bilayers. Here, the label is attached so that the NO group lies parallel to the long axis of the lipid.

Intramolecular motions and lateral diffusion of lipid through the membrane, as well as the effect of proteins and other factors on these parameters, may be observed. Quantification of effects often involves calculation of the order parameter S. If a spin probe is weakly immobilised, it shows an isotropic EPR and an order parameter of $S \approx 0.1$. In contrast, an anisotropic EPR spectrum is obtained from immobilised probes whose order parameters are at the level of $S \approx 0.7$ or higher. Spin-labelled lipids are either concentrated into one region of the bilayer or randomly incorporated into model membranes. The diffusion of spin labels allows them to come into contact with each other, which causes line-broadening in the spectrum. Labelling of phospholipids with 2,2,6,6-tetramethylpiperidine-1-oxyl (TEMPOL) is used for measurement of the flip rate of phospholipids between inner and outer surfaces as well as lateral diffusion.

Free radicals

Molecules in their triplet states (Fig. 1.7) have unpaired electrons and thus are amenable to EPR spectroscopy. Such molecules possess the property of phosphorescence, and EPR may deliver data complementary to the UV/Vis region of the spectrum. For instance, free radicals due to the triplet state of tryptophan have been observed in cataractous lenses.

Spin trapping is a process whereby an unstable free radical is being stabilised by reaction with a compound such as 5,5-dimethylpyrroline-1-oxide (DMPO). Hyperfine splittings (see Section 3.1.3) are observed that depend upon the nature of the radical.

Carcinogenesis is an area where free radicals have been implicated. While free radicals promote the generation of tumours through damage due to their high reactivity, there is, in general, a lower concentration of radicals in tumours than in normal tissue. Also, a gradient has been observed, with higher concentrations of radicals in the peripheral non-necrotic surface layers than in the inner regions of the tumour. EPR has been used to study implanted tumours in mice but also in evaluation of potential chemical carcinogens. Polycyclic hydrocarbons, such as naphthalene, anthracene and phenantrene, consist of multiple aromatic ring systems. These extended aromatic systems allow single free electrons to be accommodated and thus yield long-lived free radicals, extending the periods of time where damage can be done. Many of the precursors of these radicals exist in natural sources such as coal tar, tobacco smoke and other products of combustion, hence the environmental risk. Another source of free radicals is irradiation with UV light or γ-rays. Ozone is an oxygen radical that is present as a protective shield around the Earth, filtering the dangers of cosmic UV irradiation by complex radical chemistry. The pollution of the Earth's atmosphere with radical-forming chemicals has destroyed large parts of the ozone layer, increasing the risk of skin cancer from sun exposure. EPR can be used to study biological materials, including bone or teeth, and detect radicals formed due to exposure to high-energy radiation.

Another major application for EPR is the examination of irradiated foodstuffs for residual free radicals, and is mostly used to establish whether packed food has been irradiated.

3.2 Nuclear magnetic resonance

The essential background theory of the phenomena that allow NMR to occur have been introduced in Sections 3.1.1 and 3.1.2. However, the miniature magnets involved here are not electrons but nuclei. The specific principles, instrumentation and applications are discussed below.

3.2.1 Principles

NMR spectroscopy is a non-destructive technique that allows detection of all nuclei with an odd number of nucleons Z (the number of nucleons Z, or mass number, is the sum of the number of protons P and number of neutrons N in a nucleus: $Z = P + N$). Of particular interest to structural biology and chemistry are ^{1}H, ^{13}C, ^{15}N, ^{19}F and ^{31}P nuclei. The applications of NMR are wide, and range from elucidating the structure and quantity of small and large molecules to biomedical imaging.

Similar to other spectroscopic techniques discussed earlier, the energy input in the form of electromagnetic radiation promotes the transition of 'entities' from lower to higher energy states (Fig. 3.3). In the case of NMR,

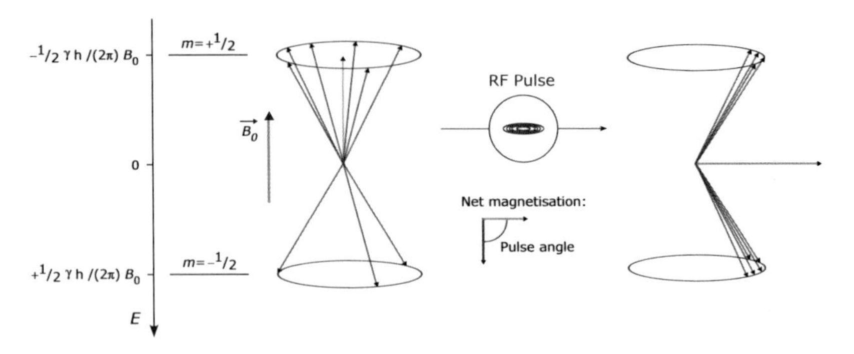

Fig. 3.3. Energy levels of a proton in the magnetic field $\vec{B}_0$. The nuclear spin of a nucleus is characterised by its magnetic quantum number m. For protons, m can only adopt $+1/2$ and $-1/2$. The corresponding energies are calculated by $-m\gamma h/(2\pi)$ B_0, where γ is a constant characteristic for a particular nucleus and h is the Planck constant. Application of a temporary magnetic field (RF pulse) causes assembly of the nuclear spins (red) into a particular direction, thus rotating the net magnetisation.

these entities are the nuclear magnetic spins, which populate energy levels according to quantum chemical rules.

In the absence of a magnetic field, the spins will be random, but when exposed to an external magnetic field $\vec{B}_0$, the nuclear spins will align with the applied field, with the spin either parallel or anti-parallel to the direction of the field. Each of the nuclear spins precesses around the axis of that field with the so-called Larmor frequency. When a temporary magnetic field is applied that oscillates at the same frequency as the precessing spins, this is called an RF pulse. This pulse gives rise to a preferred direction for the individual nuclear spins. As a result, the net magnetisation (sum of all nuclear spin vectors) is rotated by a pulse angle (Fig. 3.3).

As the energy difference between the two states in Fig. 3.3 is rather small, both states will be populated in a similar manner, resulting in a relatively small resonance effect. Increasing the magnetic field $\vec{B}_0$ increases the energy difference between the two states and thus lowers the population of the high-energy state $m = -^1/_2$. This, in turn, increases the resonance effect.

After a certain time span, the spins will return from the higher to the lower energy level, a process that is known as relaxation. The energy released during the transition of a nuclear spin from the higher to the lower energy state can be emitted as heat into the environment and in this case it is called spin–lattice relaxation. This process happens with a rate of T_1^{-1}. T_1 is termed the longitudinal relaxation time, because of the change in magnetisation of the nuclei parallel to the field. Similarly, the transverse magnetisation of the nuclei is also subject to change over time, due to interactions between different nuclei. This process is thus called spin–spin relaxation and is characterised by a transverse relaxation time T_2.

Pulse-acquire and Fourier transform methods

In 'conventional' NMR spectroscopy, the electromagnetic radiation (energy) is supplied from the source as a continuously changing frequency over a preselected spectral range (continuous wave method). The change is smooth and regular between fixed limits. During the scan, radiation of certain energy in the form of a sine wave is recorded. By using the mathematical procedure of Fourier transformation (FT), the 'time domain' can be resolved into a 'frequency domain'. For a single frequency sine wave, this procedure yields a single peak of fixed amplitude. However, because the measured signal in NMR is the re-emission of energy as the nuclei return from their high-energy into their low-energy states, the recorded radiation will decay with time, as fewer and fewer nuclei will return to the ground state. The signal measured is thus called free induction decay (FID). Alternatively, the total energy comprising all frequencies between the fixed limits can be put in all at the same time. This is achieved by irradiating the sample with a broadband pulse of all frequencies at once. The output will measure all resonance energies simultaneously and will result in a very complicated interference pattern. However, FT is able to resolve this pattern into the constituent frequencies.

Figure 3.4 depicts an FID Fourier transformed into a frequency domain spectrum of ethanol. The spectrum was acquired over eight scans (2 s acquisition each). At the start, the free induction decay signal is strong, indicating that many nuclear magnetic spins relax and return from their high-energy to the low-energy states. Note that the FID decays exponentially within the first second of the acquisition, by which time almost all spins have returned to their low-energy states. A standard proton pulse sequence is run with acquisition times of 1–4 s. The lack of a strong FID signal will have an effect on the signal-to-noise ratio of the frequency domain spectrum obtained by FT.

Because the stability and homogeneity of the magnetic field is critical for NMR spectroscopy, the magnetic flux needs to be tuned, for example by locking with deuterium resonance frequencies. The lock, which compensates for the drift of the magnetic field, is achieved by automatic adjustments that keep the lock resonance at a defined frequency. Deuterium (^{2}H, symbolised as D) is an isotope of hydrogen with a nuclear spin of 1. The use of deuterated solvents is critical for routine NMR experiments, as they do not interfere with signals from ^{1}H and ^{13}C, which

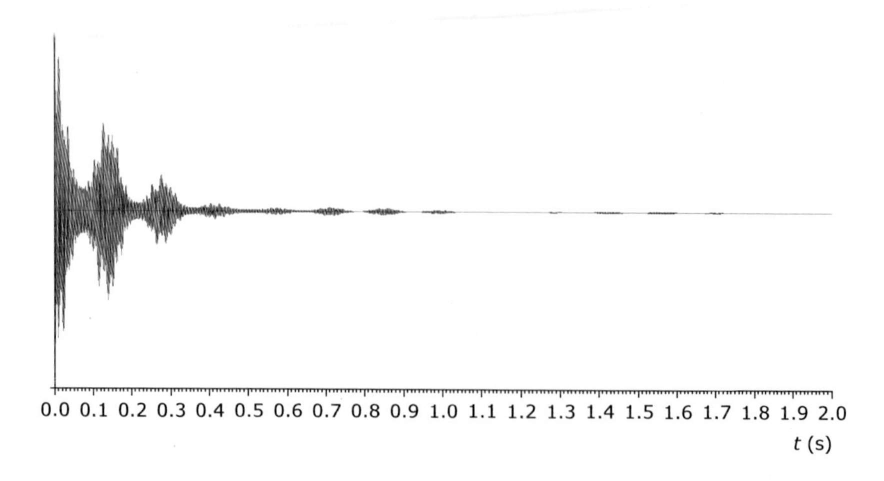

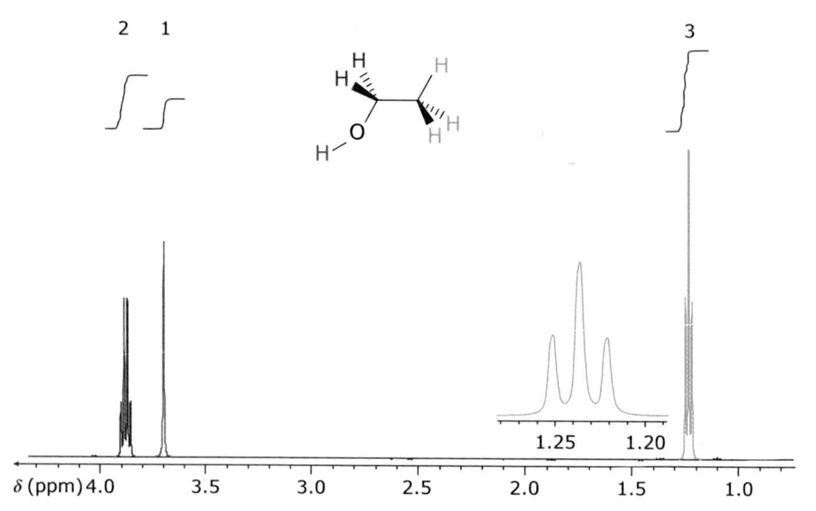

Fig. 3.4. ^{1}H NMR of ethanol recorded on a 500 MHz spectrometer in $CDCl_3$ at 30 °C. Shown is the free induction decay (top) and its frequency domain spectrum (Fourier transform methods) obtained by FT. The peak integrals reflect the number of protons eliciting a signal at a particular chemical shift.

have a nuclear spin of $^1/_2$. Usually solvents such as deuterated chloroform ($CDCl_3$), methanol (CD_3OD) and dimethyl sulfoxide (CD_3-SO-CD_3) are used for sample preparation of organic compounds. For peptides and proteins, deuterated water (D_2O) is the solvent of choice.

3.2.2 One-dimensional NMR

A one-dimensional NMR experiment measures the decay of transverse magnetisation over time. A standard proton pulse sequence is composed of three components:

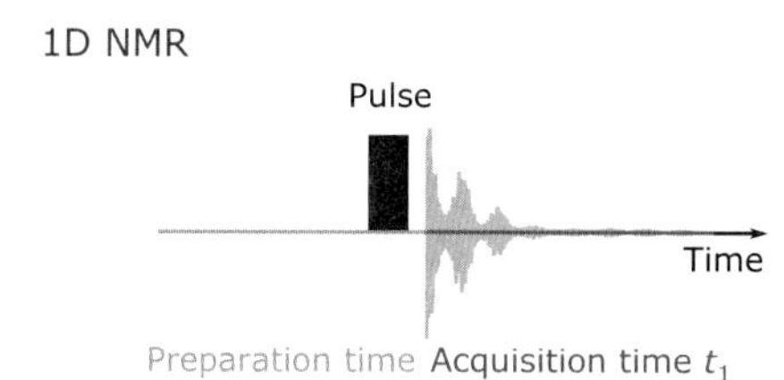

Fig. 3.5. Typical pulse sequences for one-dimensional (1D) and two-dimensional (2D) NMR. Top: Pulse sequence used for one-dimensional ^{1}H NMR. Bottom: Pulse sequence for a correlation spectroscopy (COSY) two-dimensional experiment discussed in Section 3.2.3.

- preparation or the time needed for the sample to reach equilibrium;
- the radio frequency pulse to disturb the equilibrium; and
- the acquisition time (t_1) during which the FID signal is recorded.

In order to improve the signal-to-noise ratio, the experiment is repeated in successive scans and the data are summed up prior to Fourier transformation. A typical ^{1}H NMR pulse sequence is depicted in Fig. 3.5.

Chemical shift

The molecular environment of a proton governs the value of the applied external field at which the nucleus resonates. This is recorded as the chemical shift (δ) and is measured relative to an internal standard, which for most non-aqueous sample solutions is tetramethyl silane (TMS; $Si(CH_3)_4$) because it contains 12 identical protons and thus produces a single peak. As TMS is only soluble in organic solvents, sodium salts of 2,2-dimethyl-2-silapentane-5-sulfonate (DSS) are used in aqueous samples. The chemical shift arises from the applied field inducing secondary fields of about 0.15–0.2 mT at the proton by interacting with the adjacent bonding electrons:

- If the induced field opposes the applied field, the latter will have to be at a slightly higher value for resonance to occur. The nucleus is said to be shielded, the magnitude of the shielding being proportional to the

electron-withdrawing power of electronegative substituents, π-bonds and hydrogen bonding.
- Alternatively, if the induced and applied fields are aligned, the latter is required to be at a lower value for resonance. The nucleus is then said to be deshielded.

In an NMR spectrum, the chemical shift δ is plotted along the x-axis, and measured in parts per million (ppm) instead of the actual magnetic field strength B. This conversion makes the recorded spectrum independent of the particular magnetic field strength used; the signal of the internal standard TMS appears at $\delta = 0$ ppm, as it is deshielded relative to most other nuclear environments. The type of proton giving rise to a particular band may thus be identified by the resonance peak position, i.e. its chemical shift. A summary of typical ^{1}H NMR chemical shifts of common functional groups is presented in Fig. 3.6.

Multiplicity

Multiplicity or coupling is caused by the effect of neighbouring atoms, as the proximity of n vicinal hydrogen atoms will cause the signal to be split into $(n + 1)$ lines. An NMR signal with no vicinal hydrogens (i.e. one isolated hydrogen) will appear as a single line or a singlet, one vicinal hydrogen will give rise to a doublet, two to a triplet and so on. For example, in the ethanol ^{1}H NMR spectrum depicted in Fig. 3.4, two signals at 1.16 and 3.61 ppm appear as a triplet ($C\underline{H_3}CH_2OH$) and a quartet ($CH_3C\underline{H_2}OH$), respectively. Multiplicity is a very useful feature of NMR spectra, as it provides an elegant insight into the chemical environment for each atom eliciting a signal, as well as its directly bonded neighbours.

J-coupling (also called scalar or spin–spin coupling) is an extension of the concept of multiplicity, with the magnitude of the coupling constant J being a measure of an interaction between two coupling protons. This gives rise to the splitting of the three bands into several finer bands (hyperfine splitting). The hyperfine splitting yields valuable information about the near-neighbour environment of a nucleus.

Signal intensity

Taken as the area under the peak, the signal intensity is proportional to the number of nuclei giving rise to the signal. An integral represents the relative ratio of the number of protons for each resonance. The concept

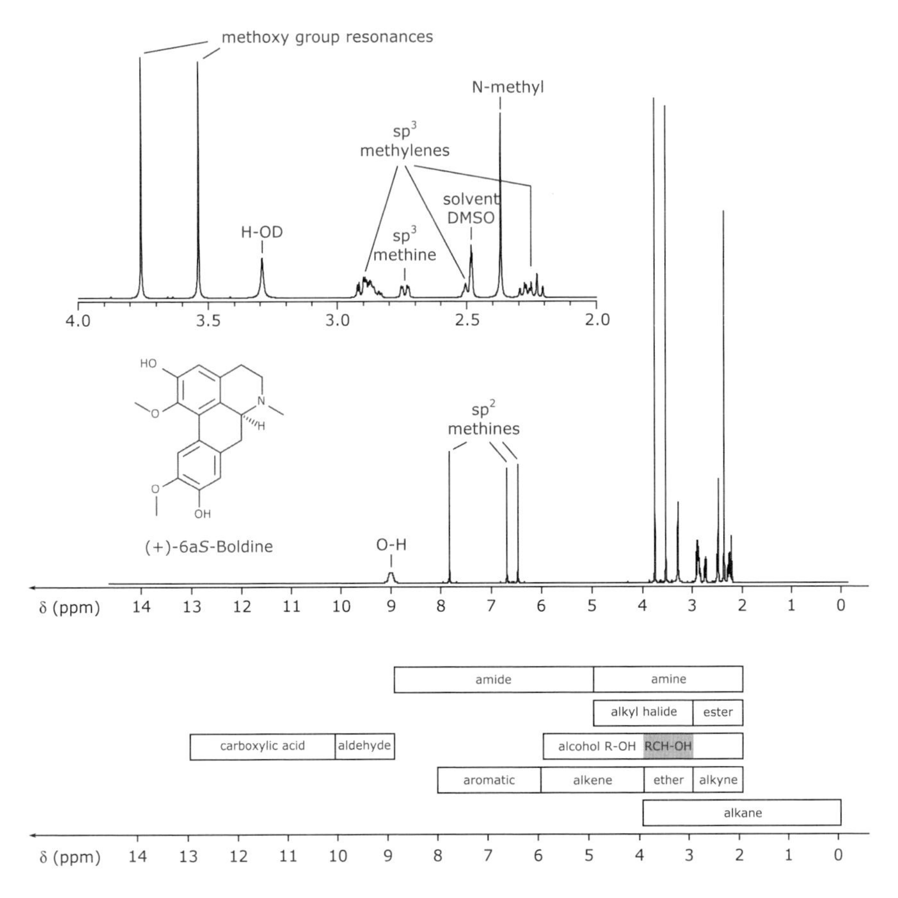

Fig. 3.6. Overview of ^{1}H NMR chemical shifts for common chemical groups, illustrated using the natural product boldine. Top: ^{1}H NMR spectrum of (+)-6a*S*-boldine recorded on a 600 MHz spectrometer at 30 °C in DMSO-d$_6$. Peaks are assigned with the nuclei causing the observed resonances. Inset: Expansion of the spectrum in the region $\delta = 2$–4 ppm. Bottom: ^{1}H chemical shift regions for common chemical groups. The H-OD peak is due to residual water (H_2O) in the sample in which one of the protons has been exchanged with deuterium.

is exemplified in the ethanol ^{1}H NMR spectrum depicted in Fig. 3.4, where the three peaks for CH_3CH_2OH are integrated and show the proportions of 3:2:1 for the methyl, methylene and hydroxyl protons respectively.

Hydrogen–deuterium exchange

Covalently bound hydrogen atoms in alcohols or amides can be replaced by a deuterium atom from the solvent in a reaction called hydrogen–deuterium (H/D) exchange (see for example H-OD in Fig. 3.6). H/D exchange is of particular importance in protein NMR, where amide hydrogen atoms of the protein peptide bond are exchanged against deuterium from the solvent. As the exchange reaction is effectively an acid–base reaction, it is strongly pH dependent.

Nuclear Overhauser effect

It was mentioned above that nuclear spins generate magnetic fields that can exert effects through space, for example as observed in spin–spin coupling. This coupling is mediated through chemical bonds connecting the two coupling spins. However, magnetic nuclear spins can also exert effects in their proximal neighbourhood via dipolar interactions. The effects encountered in the dipolar interaction are transmitted through space over a limited distance of the order of 5 Å (0.5 nm) or less. Such an interaction can lead to the nuclear Overhauser effect (NOE), observed as the changing signal intensity of a resonance when the state of a near neighbour is perturbed from equilibrium. Because of the spatial constraint, this information enables conclusions about the three-dimensional geometry of the molecule being examined.

Summary of NMR parameters

The parameters derived from NMR spectra that can be used to derive structural constraints of a small molecule or protein are summarised in Table 3.1.

Table 3.1. NMR-derived structural parameters of molecules.

Parameter	Information	Example/comment
Chemical shift δ	Chemical group Secondary structure	^{1}H, ^{13}C, ^{15}N, ^{31}P
J-couplings (through bond)	Dihedral angles	^{3}J(amide-H,Hα), ^{3}J(Hα, Hβ), etc.
NOE (through space)	Interatomic distances	<0.5 nm
Solvent exchange	Hydrogen bonds	Hydrogen-bonded amide protons are protected from H/D exchange, while the signals of other amides disappear quickly
Relaxation/line widths	Mobility, dynamics, conformational/ chemical exchange Torsion angles	The exchange between two conformations, but also chemical exchange gives rise to two distinct signals for a particular spin
Residual dipolar coupling	Torsion angles	^{1}H–^{15}N, ^{1}H–^{13}C, ^{13}C–^{13}C, etc.

^{13}C NMR spectroscopy

Carbon NMR is based on the isotope ^{13}C, which shows only 1.1% natural abundance. As such, ^{13}C NMR is many orders of magnitude less sensitive compared with ^{1}H NMR without isotopic enrichment. It is, however, invaluable for the structure elucidation of biologically relevant molecules. The chemical shifts for different carbon environments in a molecule are in the range of 0–220 ppm and are governed by the same principles discussed for ^{1}H NMR above. Deshielding effects of heteroatoms and π-bonds will cause the resonance to appear down field, while shielded resonances appear further up field. Modern ^{13}C NMR spectroscopy uses broadband proton decoupling, which gives rise to a single line for each resonance in the absence of ^{13}C–^{1}H decoupling. Decoupling is achieved with a saturation pulse and removes the spin–spin coupling – in this case between ^{13}C and ^{1}H nuclei. ^{13}C spectra are thus much simpler and cleaner when compared with ^{1}H spectra. The main disadvantage, however, is the fact that multiplicities in these spectra cannot be observed, i.e. it cannot be decided whether a particular ^{13}C is associated with a methyl (CH_3), a methylene (CH_2) or a methine (CH) group. Some of this information can be regained by irradiating with an off-resonance frequency during a decoupling experiment. Another method is called distortionless enhancement by polarisation transfer (DEPT), where sequences of multiple pulses are used to excite nuclear spins at different angles, usually 45, 90 or 135°. Although interactions have been decoupled, in this situation the resonances exhibit positive or negative signal intensities, dependent on the number of protons bonded to the carbon. When exciting at 135° (DEPT-135), for example, a methylene group yields a negative intensity, while methyl and methine groups yield positive signals. A modern variation of this pulse sequence is the two-dimensional phase-sensitive heteronuclear single quantum correlation (HSQC) experiment.

3.2.3 Two-dimensional NMR

As we have seen above, the observable in pulse-acquired Fourier transform NMR is the decay of the transverse magnetisation, called free induction decay (FID). The detected signal is thus a function of the acquisition (detection) time t_2 (see Fig. 3.5). Within the pulse sequence, the time t_1 (evolution time) describes the time between the first pulse and signal detection. If t_1 is systematically varied, the detected signal becomes a function of both t_1 and t_2, and its Fourier transform comprises two frequency components. The two components form the basis of a two-dimensional spectrum.

Table 3.2. A summary of the most commonly used two-dimensional experiments.

Experiment		Applications
Homonuclear two-dimensional		
COSY	Correlation spectroscopy	Establishes connection between directly coupled spins
TOCSY	Total correlation spectroscopy	Establishes correlation between all spins Through-bond and -space connectivity
NOESY	Nuclear Overhauser effect spectroscopy	Through-space connectivity via dipolar coupling using longitudinal cross-relaxation
ROESY	Rotating Overhauser effect spectroscopy	Through-space connectivity via dipolar coupling using transverse cross-relaxation
Heteronuclear two-dimensional		
HSQC	Heteronuclear single quantum correlation	Single bond heteronuclear connectivity, direct (^1J) heteronuclear assignment
HMQC (HMBC)	Heteronuclear multiple quantum correlation	Multiple bond heteronuclear connectivity, long-range (^2J, ^3J and ^4J) heteronuclear assignment

Correlated two-dimensional NMR spectra show chemical shifts on both axes, and couplings between particular nuclei appear as off-diagonal signals (Fig. 3.7). Using different pulse sequences allows monitoring of different homonuclear and heteronuclear couplings. A summary of the most commonly used two-dimensional experiments and their applications is given in Table 3.2.

3.2.4 Instrumentation

Schematically, an analytical NMR instrument is very similar to an EPR instrument, except that instead of a klystron generating microwaves two sets of coils are used to generate and detect radio frequencies (Fig. 3.8). Samples in solution are contained in sealed tubes, which are rotated rapidly in the cavity to eliminate irregularities and imperfections in sample distribution. In this way, an average and uniform signal is reflected to the receiver to be processed and recorded. In solid samples, the number of spin–spin interactions is greatly enhanced due to intermolecular interactions absent in liquid samples due to translation and rotation movements. As a result, the resonance signals broaden significantly. However, high-resolution spectra can be obtained by spinning the solid sample at an angle of 54.7° (magic angle spinning). The sophisticated pulse sequences necessary for multi-dimensional NMR require a certain geometric layout of the radiofrequency coils and sophisticated electronics. Advanced computer facilities are needed for operation of NMR instruments, as well as analysis of the acquired spectra.

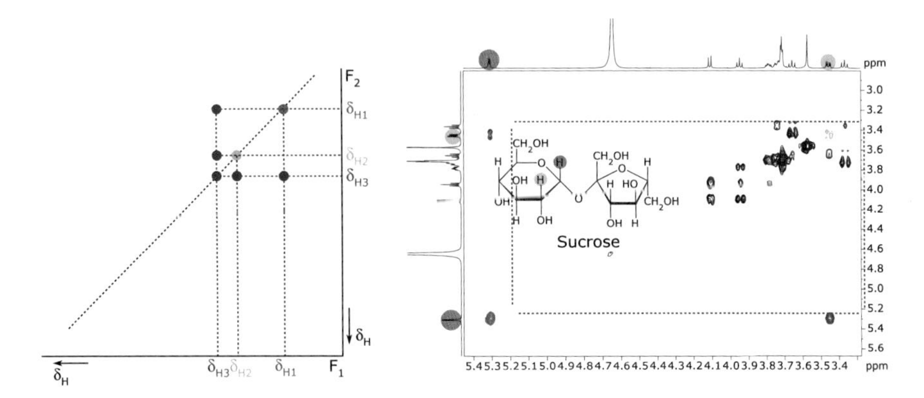

Fig. 3.7. Left: Schematics of a correlated two-dimensional ^{1}H NMR spectrum. H3 couples with H1 and H2. H1 and H2 show no coupling. Right: A two-dimensional correlation spectroscopy (COSY) spectrum of sucrose in D_2O acquired on a 600 MHz spectrometer at 30 °C.

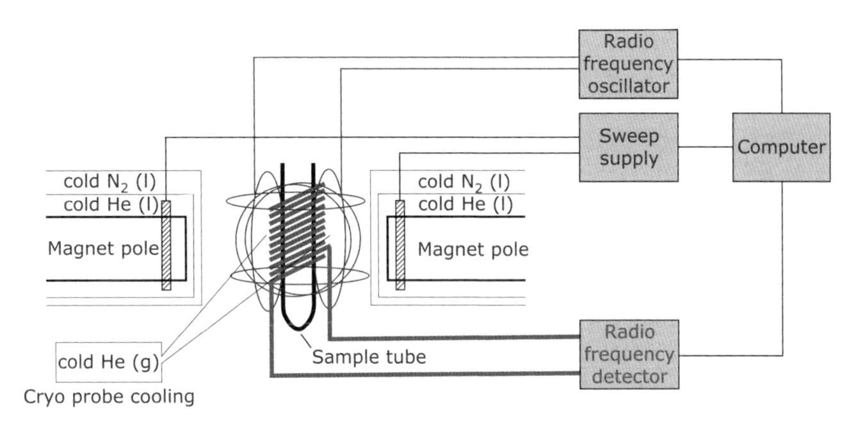

Fig. 3.8. Schematic diagram of an NMR spectrometer with cryo-probe. The probe coil is held at 25 K by circulating cold helium gas. Cryo-probes allow a significant increase of the sensitivity compared to room temperature probes. The magnet needs to be cooled to liquid helium temperature (4 K) to achieve the state of superconductance.

The frequency of the external magnetic field is most commonly reported as the resonance frequency of the spin of the proton in MHz or GHz; alternatively, the strength of the external magnetic field is reported in Tesla (T). The largest magnet in use today is operating at 1 GHz (23.5 T); standard NMR spectrometers for academic and industrial use range from 300 MHz (7.0 T) to 900 MHz (21.2 T). Magnetic resonance imaging, in contrast, is conducted with magnetic fields in the range of 1–3 T, as humans, animals or generally living organisms are being exposed (see Section 3.2.5).

3.2.5 Applications

Structure determination of small molecules

NMR is a powerful method for the structure elucidation of small molecules. A proton experiment can be used to confirm the structural identity of a compound, quantify the amount of an analyte in a mixture and track the progress of a synthetic reaction. Heteronuclear two-dimensional experiments can be used to elucidate the structure of unknown compounds and have allowed the field of natural products chemistry to flourish over the last 50 years. Moreover, *J*-based configurational analysis allows the determination of the spatial orientation of substituents in a molecule, a structural technique second only to X-ray diffraction crystallography. The sensitivity of instrumentation requiring between 1 and 20 mg for complete characterisation, ease of performing sophisticated pulse sequences using modern instrumentation and the applicability of this technique to analyse all compounds containing ^{1}H, ^{13}C, ^{15}N, ^{19}F and ^{31}P nuclei make NMR spectrometers an irreplaceable resource in the study of the structure of small molecules.

Structural determination of macromolecules

The structures of proteins up to a mass of about 50 kDa can be determined with biomolecular NMR spectroscopy. The development of magnets with very high field strengths (currently 1 GHz), allowing increased sensitivity and dispersion of signal, continues to push the size limit of macromolecules that can be investigated with this technique. As NMR-based study of proteins relies on two- and three-dimensional experiments involving low-abundance ^{13}C and ^{15}N nuclei, it is common to use expression of recombinant proteins and isotopic labelling. The introduction of cryo-probe technology has reduced the amount of sample required for NMR from millimolar to micromolar concentrations. Heteronuclear multidimensional NMR spectra need to be recorded for the assignment of all chemical shifts (^{1}H, ^{13}C, ^{15}N). The data acquisition can take several weeks, after which spectra are processed (by FT) and improved with respect to digital resolution and signal-to-noise ratio. Assignment of chemical shifts and interatomic distances is carried out with the help of computer programs. All experimentally derived parameters are then used as restraints in a molecular dynamics structure calculation. The result of a protein NMR structure is an ensemble of structures, all of which are consistent with the experimentally determined restraints but converge to the same fold. For databases of structures derived by NMR methods see Section 3.4.6.

Ligand binding and drug discovery

NMR-based screening is a well-established tool in drug discovery and is applicable to a wide variety of target proteins. The technique is able to detect a small molecule (ligand) binding to the protein (target) and relies on two main approaches, target-based screening and ligand-based screening:

- In target-based screening, the chemical shifts of the target protein are studied in the presence and absence of a ligand. A protein is incubated with a small molecule and binding is confirmed upon changes in the chemical shift of the macromolecule via two-dimensional ^{1}H–^{13}C and ^{1}H–^{15}N NMR experiments. In addition to verification of binding ('hit validation'), the target-based NMR approach can provide information on the nature of the binding interaction such as location of the binding site.
- Ligand-based screening techniques such as WaterLOGSY and saturation transfer difference (STD) are based on the detection of the difference of ligand signal in the presence and absence of a protein. In STD NMR (Fig. 3.9), the difference between selective saturation of different regions of two spectra are noted: that of protein (on-resonance saturation) and a region of the spectrum where no signals are present (off-resonance saturation). The difference is then calculated between the on-resonance and off-resonance spectra. If a ligand is bound to the protein, the resonances for hydrogen atoms involved in binding will receive magnetisation transfer via the nuclear Overhauser effect from the protein and will therefore appear enhanced in the difference spectrum, while the free ligand resonances will be absent. As STD NMR is a ligand-based detection method, its advantages are short acquisition times and the fact that it can be performed on medium-field magnets. Because the protein spectrum itself is not analysed in this technique, it is applicable to any target protein (i.e. no mass restriction), and there is no requirement for high-field magnets. Due to ease of implementation and its applicability across a range of protein targets, STD NMR is one of the most commonly used NMR-based screening methods in drug discovery today.

Magnetic resonance imaging (MRI)

The basic principles of NMR can be applied to imaging of live samples. The resonance frequency of a particular nucleus is proportional to the strength of the applied external magnetic field. If an external magnetic field gradient is applied, then a range of resonant frequencies are observed,

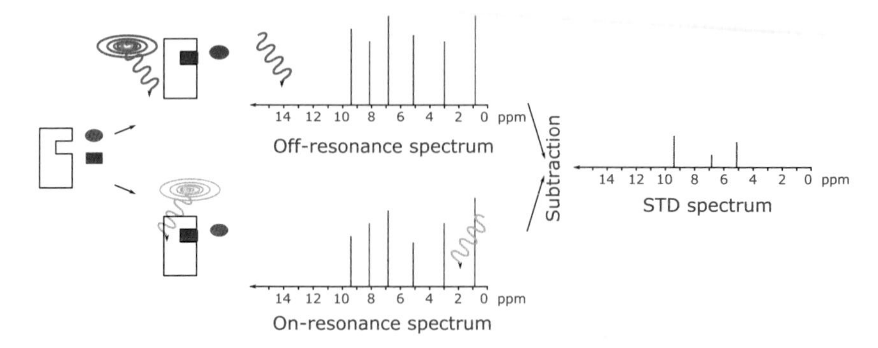

Fig. 3.9. Principle of STD experiments using ^1H NMR. The experiment is based on the transfer of magnetisation from the protein to the bound ligand. It involves subtracting the on-resonance spectrum (where the frequency of the saturating pulse is chosen for a region that contains only the protein-specific resonances, usually between 0 and 2 ppm; yellow) from the off-resonance spectrum (where the frequency of the saturating pulse targets a region with no ligand- or protein-specific resonances; green). The difference spectrum will then only show peaks arising from a protein-bound ligand, which can subsequently be identified through comparison with reference spectra.

reflecting the spatial distribution of the spinning nuclei. The number of spins in a particular defined spatial region gives rise to the spin density as an observable parameter. This measure can be combined with analysis of the principal relaxation times (T_1 and T_2).

Magnetic resonance imaging makes use of these phenomena and can be applied to large volumes in whole living organisms. The technique has a central role in routine clinical imaging of large-volume soft tissues. Because the proton is one of the more sensitive nuclei and abundant in all biological systems, ^1H resonance is used almost exclusively in the clinical environment. Thus, MRI is used to monitor water as the most ubiquitous fluid in biological cells. Water is distributed differently in different tissues, but constitutes about 55% of body mass in the average human. In soft tissues, the water distribution varies between 60 and 90%. In addition to imaging the location of water in tissue, it is also possible to image the flux, as either bulk flow or localised diffusion.

In terms of whole-body scanners, the entire picture is reconstructed from images generated in contiguous slices. Thus, MRI can be applied to the whole body or used for specific organ investigations on head, thorax, abdomen, liver, pancreas, kidney and skeletomuscular regions. The use of contrast agents with paramagnetic properties has enabled investigation of organ function, as well as blood flow, tissue perfusion, transport across the blood–brain barrier and vascular anatomy. Resolution and image contrast are major considerations for the technique and are subject to continuing development. The resolution depends on the strength of the magnetic field and the availability of labels that yield high signal strengths. The instruments used for MRI for clinical imaging typically operate with field strengths of up to 3 T, but experimental instruments can operate at more than 20 T, allowing the imaging of whole live organisms with almost enough spatial and temporal resolution to follow regenerative processes continuously at the single-cell level. Equipment cost and data acquisition time remain important issues. On the other hand, MRI has no known adverse effects on human health and thus provides a valuable diagnostic tool, especially due to the absence of the hazards of ionising radiation.

3.3 Electron microscopy

The resolving power of optical systems, a concept developed mainly by Lord Rayleigh (1842–1919) and Ernst Abbe (1840–1905), defines the ability of an imaging device to separate points that are located at a small distance. The minimum resolvable distance is often called (angular) resolution. The Rayleigh criterion holds that two point-like light sources are just resolved when the main diffraction maximum of one image coincides with the first minimum of the other. Any greater distance between the two point sources means they are well resolved; any smaller distance makes the two points non-resolvable. In terms of a circular aperture, this leads to the equation

$$\sin\theta = 1.22 \times \frac{\lambda}{D}, \tag{3.2}$$

where θ is the angular resolution, λ the wavelength of light used to image and D the diameter of the aperture of the lens system; the factor of 1.22 is derived from the Bessel function of first-order first kind, divided by π ($J_1/\pi = 1.22$).

In the early 1930s, Ernst Ruska and Max Knoll built the first electron microscope, reasoning that an electron beam with wavelengths much lower than that of visible light might have a much greater resolving power. Electron microscopes were slow to develop, but a major advance in the biological sciences came about in the 1960s with the discovery of fixation methods and solvent-miscible resins, which enabled the samples to be prepared for electron microscopy. Current electron microscopes can work at physical resolution limits of 1–2 Å, i.e. at atomic resolution. In 2008, an electron microscope with a resolution of 0.5 Å was built (a transmission electron aberration-correlated microscope), but recent research indicates that further increases in resolution are coming up against physical limits (Uhlemann *et al.*, 2013).

3.3.1 Principles

Electrons possess properties of particles as well as electromagnetic waves (de Broglie hypothesis). As electron microscopes work by passing a focused electron beam through samples or shining the beam onto samples, an inherent problem is that the beam is susceptible to all matter. Therefore, within the microscope, the electron beam needs to operate for much of its path under conditions of high vacuum. Secondly, the penetration depth of an electron beam through samples is rather low, as they are deflected, reflected, absorbed and otherwise lost from the sample. Samples with a large proportion of elements with high atomic number (Z) are more likely to interact with the incident electron beam because they possess larger numbers of electrons surrounding the atomic nucleus and there is thus a greater potential for collisions to occur. Likewise, thicker samples are more likely to suffer collision between incoming electrons and sample, because of the greater likelihood of electrons interacting with multiple stacked atoms as they pass into the sample.

The interaction of electrons of the incident beam with matter in the sample gives rise to a series of radiation phenomena and spectra, all of which can be used to provide information about the sample (Fig. 3.10). The observed effects include the directly transmitted electrons, elastically scattered electrons, inelastically scattered electrons, Bremsstrahlung X-rays, infrared, visible and UV light rays, characteristic X-rays, Auger electrons, back-scattered electrons and secondary electrons. Of these, the back-scattered electrons and secondary electrons are more typically associated with scanning electron microscopy and will be considered in Section 3.3.3.

Another aspect to consider is the effect of contrast, which is required by the human eye to form an image. As electron waves are not part of the visible spectrum, they cannot be used to generate colour, which is a means for human eyes to distinguish incident light of different wavelengths. Images therefore need to be constructed by detecting electrons and converting detection frequencies into light waves or electrical signals that can be visualised.

Formation of a visible image from the interaction of an electron beam with the sample is complex and has given rise to the major technical requirements for electron microscopes: high vacuum, which is not conducive to life, and ultrathin samples.

The major limiting factor for the resolution in electron micrographs is not the resolving power but rather the contrast. For example, the resolving power of an electron microscope may be in the order of 2–3 Å, yet for most

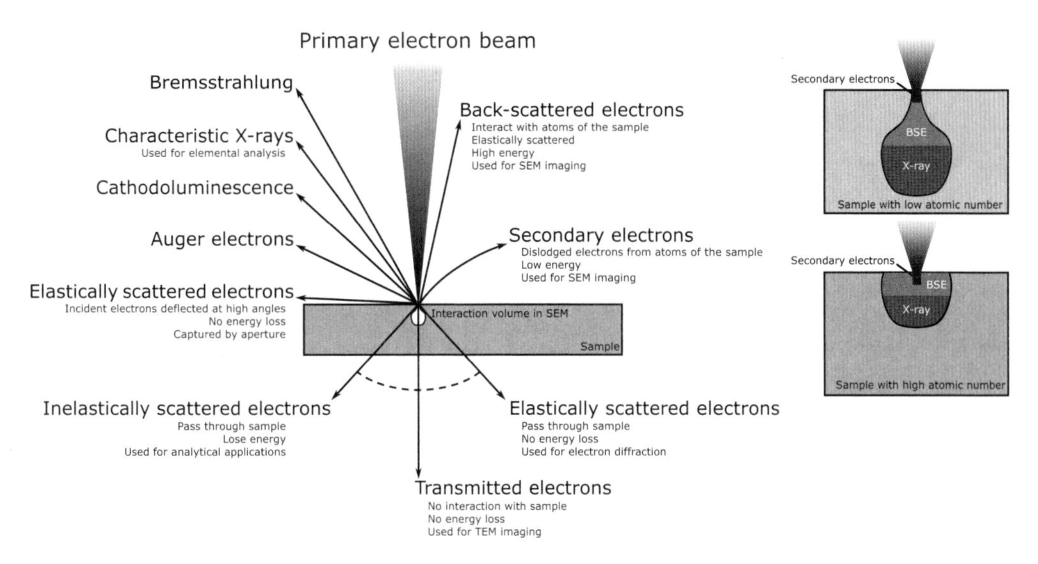

Fig. 3.10. Interactions of an electron beam with a specimen. Left: Types of electrons and electromagnetic radiation observed when an electron beam hits a sample. Right: Interaction volume of an electron beam with a sample in scanning electron microscopy (SEM). The same sort of interaction would occur in transmission electron microscopy (TEM); however, as the samples are much thinner in TEM, electrons penetrating deeper into the sample emerge again from the bottom of the section. BSE, back-scattered electrons.

biological specimens, resolution is limited to about 10–50 Å (1–5 nm). The contrast formation in an electron microscope depends on the mode of operation. The most common way is simply by absorption of electrons in the sample (bright-field imaging mode), but diffraction (see Section 3.3.6) and electron energy loss (see Sections 3.3.2 and 3.3.5) can also be used. There are two fundamentally different contrast phenomena (Fig. 3.11):

- Amplitude contrast results from variation in mass or thickness of the specimen.
- For phase contrast, one exploits the wave nature of electrons; more specifically, it is the differences in the phase of electron waves scattered through the specimen that allow contrast formation.

In many cases, both phenomena contribute to the formation of an image.

3.3.2 Transmission electron microscopy (TEM)

In TEM, an electron beam is directed through a sample that is held before the objective lens (Fig. 3.12). Electrons transmitted through the sample are

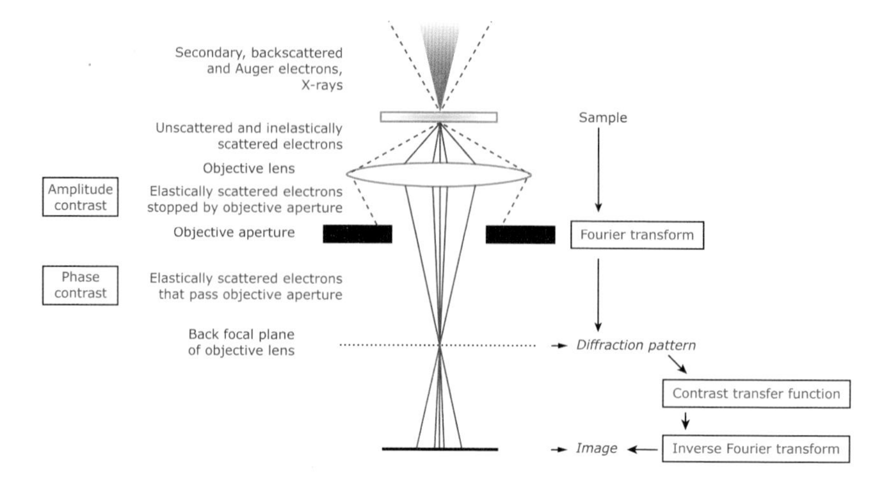

Fig. 3.11. Contrast in electron microscopy. The schematic on the left shows where amplitude and phase contrast arise in the electron microscope. The back focal plane of the objective lens is important for collecting a diffraction image in electron crystallography (see Section 3.3.6). The flow chart on the right shows the mathematical links between the three-dimensional sample, its diffraction pattern and the image in electron microscopy.

used to form the TEM image. Directly transmitted electrons are those that pass through the sample without interacting with any atom. Therefore, they pass through the sample non-deflected and without loss of energy. After exiting the sample, they progress unimpeded through the microscope lenses to the image-forming devices. As the electrons cannot penetrate far through a sample, very thin specimens are required for TEM studies, typically no more than 60–100 nm in thickness. Thicker specimens can be used if the electrons are accelerated with a higher-than-standard voltage.

Some electrons collide with an atomic nucleus on their passage through the sample. These are scattered elastically, i.e. they are deflected at high angles but without loss of energy. This can be envisioned like a billiard ball cannoning off another ball, except that the second ball does not move. These elastically scattered electrons end up colliding with an aperture placed just beneath the sample and are thus removed from the transmitted beam, i.e. do not contribute to image formation.

Inelastically scattered electrons arise from the collision of incident electrons with electrons in the sample. Some of the energy of the incident electron is imparted onto the sample. These electrons pass through the sample with lower energy and are not filtered by the aperture. The energy

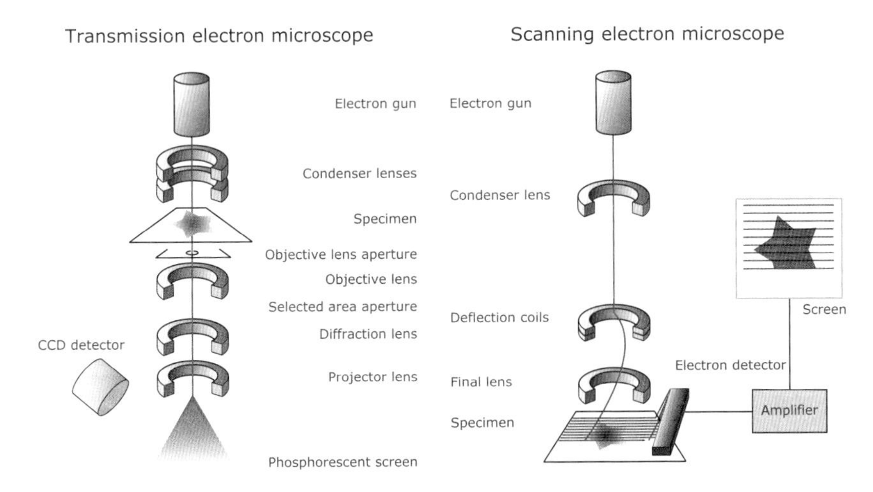

Fig. 3.12. Schematic comparison of a transmission and a scanning electron microscope.

loss is correlated with the atomic number *Z* of the atoms with which the electrons collide. Inelastically scattered electrons are used in analytical areas of microscopy, particularly in electron energy loss spectroscopy (EELS), a method that can be used to map specific elements in a sample.

Formation of a TEM image is thus composed of two phenomena. Directly transmitted electrons are captured by the image-forming device and converted into visible light through the process of phosphorescence (see Section 2.3.1). Elastically scattered electrons are filtered by the objective lens aperture and are thus not available to be focused by the objective lens and therefore subsequently not projected onto the image-forming device. Contrast in TEM is observed as the different signal intensities of certain regions of the image-forming device. Unscattered incident electrons that are not passing through the sample, yet reach the detector (i.e. electrons that are on a trajectory through areas where there is no sample) give rise to bright zones. As one increases the magnification, the beam is thus focused on regions of interest.

With biological samples, which generally consist of atoms with low atomic number, it is very difficult to provide contrast in TEM. This contrast can be achieved by incorporating elements into the sample with high atomic number, usually introduced as fixatives or stains. Osmium tetroxide (OsO_4), for example, has an unusual predilection for membranes; it tends to align along membranes in fixed samples. The interaction of electrons with osmium-enriched membranes thus leads to enhanced elastic scattering of electrons at the site of membranes due to the high atomic number (and thus large number of electrons) of osmium. Membranes therefore appear dark in a TEM image when stained with OsO_4.

3.3.3 Scanning electron microscopy (SEM)

Scanning electron microscopy is used predominantly to form images of surfaces. For this reason, SEM images are more readily interpreted by many people, even those who have no specific training in structural or cellular biology, or in electron optics. Indeed, as it is possible to colour the images artificially based on any chosen characteristic, SEM images have featured in many forms of art and literature.

Scanning electron microscopes, like transmission electron microscopes, are high-vacuum instruments that contain an electron gun and a series of electromagnetic lenses that focus the beam into a convergent beam called a probe (Fig. 3.12). The probe diameter can vary between 1 nm and 1 μm, and directly influences the resolving power of the microscope and hence the achievable magnification. The electron current reaching the specimen (probe current) can be adjusted using the condenser (Fig. 3.12) and is varied in order to improve the signal-to-noise ratio. A deleterious effect of increasing probe current, however, is the potential of damage to the specimen. Much of the skill in effectively operating a scanning electron microscope therefore lies in understanding this and other parameters of operation.

The lenses in a scanning electron microscope work to form a focused beam. Scanning coils (hence the name of the microscope) placed within the final lens raster the electron probe across the surface of the sample. During the rastering operation, the coils drag the probe in a series of highly precise lines, so that the electron beam is projected onto the sample as a spot, constantly moving in a series of parallel lines across the surface. At each point, the beam penetrates into the sample and the electrons will interact with atoms in the sample and elicit effects as described above for TEM. The three-dimensional zone that the electrons tunnel through in the sample is called the interaction volume (see Fig. 3.10). In typical biological samples, the interaction volume has the shape of a pear; in samples with high atomic number (Z), the interaction volume is shaped more like a hemisphere.

The incident electrons can be scattered in any direction, but because the samples in SEM are generally very thick, only those that are directed back to the sample surface and beyond into the specimen chamber are used for imaging. Depending on the interaction, there are two sorts of emergent electrons:

- Firstly, an electron that enters a sample may interact with a valence electron in the sample. As much of the energy of the incident electron is transferred to the spinning electron, this is called an inelastic interaction, dislodging the valence electron. As the dislodged electron will emerge from the sample, usually at low energy (<50 eV), it is now called the secondary electron. The number of secondary electrons generated in a beam–sample interaction is largely independent of the atomic number Z of the atoms in the sample.
- Other electrons of the incident beam will interact with atomic nuclei of the sample and are scattered elastically, i.e. with very little loss of energy (loss usually <1 eV). They possess much higher energies than secondary electrons and about the same energy with which they entered the sample. As these electrons can be scattered through angles of up to 180°, they can be bounced back out of the sample and are thus called back-scattered electrons. The number as well as the energy of the back-scattered electrons is highly dependent on the atomic number Z of atoms in the sample.

The two types of electrons emerge from the sample at the mouth of the interaction volume in the split second the electron beam has hovered above it on its path. For detection, these electrons are captured and projected against a scintillator to form light photons that pass along a photomultiplier tube to be projected onto a cathode ray tube or a light-sensitive sensor such as a charge-coupled device (CCD) camera. In cathode ray tubes, scanning coils are synchronised with the deflection coils (Fig. 3.12). Each point on the sample has a conjugate point on the cathode ray tube, so that electrons emerging from a defined point of the sample end up being projected on to a defined point. Detectors for back-scattered electrons are designed to capture electrons of relatively high energy; those for secondary electrons capture low-energy electrons.

Above, we saw that in TEM one observes electrons that are transmitted through the sample. However, the transmission electron microscope can also be used as a scanning microscope, if scanning coils are fitted to the condenser lens of the transmission electron microscope just above the sample and the beam is rastered over the specimen. This hybrid microscopy technique is called a scanning transmission electron microscopy (STEM) and is used for procedures that require the high-voltage properties of TEM while providing the imaging and analytic capabilities of SEM.

3.3.4 Applications

Morphological assessments of cellular structure

One of the primary contributions of both TEM and SEM has been in the description of morphology of cellular structures (Fig. 3.13). Electron microscopy has been applied in countless studies to image tissues, cells, microbes, viruses – in fact all forms of life at the cellular level. The technique has been a major contributor to our understanding of structure and function of cells, their organelles and components in normal and pathological conditions. Many cell features, hinted from earlier studies by light microscopy, came into stark realisation using TEM. For example, Golgi bodies had first been seen by Camillo Golgi some 40 years before electron microscopes were invented but are among the ubiquitous and heavily studied components of cells. Other organelles, such as the plasma membrane-associated caveolae only came to light after the advent of electron microscopy.

An area where electron microscopy continues to be applied routinely is in human pathology testing. Assessment using TEM forms a major part of determining causes of renal failure and diseases of the filtration unit of the kidney, the glomerulus. This structure is subject to numerous pathological changes, because of diabetes and autoimmune diseases. Based on structural alterations of the cells and basement membranes in the glomerulus, an experienced pathologist can readily determine the cause of kidney dysfunction. Similarly, the range of conditions known as cilial dyskinesia are genetic disorders characterised by dysfunction of the cilia that line the respiratory system, auditory system and female reproductive tract and that are present in male sperm. The rare genetic condition arises from the inability of some of the structural components of cilia to form properly.

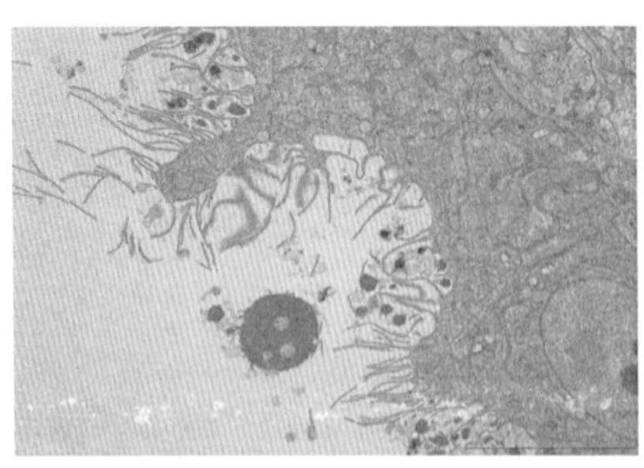

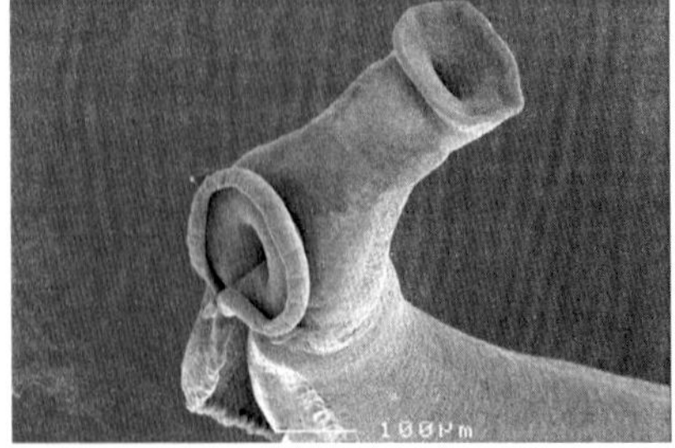

Fig. 3.13. Left: Transmission electron micrograph of the intestinal region of a female schistosome. The dark circular region is a partially digested red blood cell. Right: Scanning electron micrograph of the anterior region of a male schistosome.

Although genetic testing for the condition is feasible, TEM analyses for defects can rapidly aid in diagnosis of the disorder.

For most human tissues, we have a good inventory of cellular structure; however, the discovery of structural components is ongoing, particularly in dissecting new adaptations of incompletely known organisms, including many microbial organisms and many unicellular eukaryotes.

Advanced ultrastructural methods

Transmission electron microscopes typically project two-dimensional images, but there are many situations where more information can be gained from a three-dimensional image. In recent years, substantial effort has been put into techniques that map structures in three dimensions. This methodology is called tomography and began in the 1960s but had to await the advent of digital imaging, automated control of microscopes and fast computing to be fully realised.

In electron tomography, a specimen of 150–300 nm thickness is imaged by TEM and then tilted from the horizontal plane through different angles, with electron micrographs being taken at each point. The images are then stacked and processed to remove aberrations that arise because of specimen drift and compression. Using a method called filtered back projection, a three-dimensional image is constructed from the two-dimensional sections. This complex three-dimensional image can then be sliced in different planes, and coloured to highlight membranes and other structures to provide advanced information on the structure of cells. When used in conjunction with rapid freezing techniques, electron tomography becomes a powerful tool to model dynamic interactions of organelles and structures in cells.

A new tool for three-dimensional viewing is the ingenious serial block-face scanning electron microscopy (SBEM). This method combines TEM specimen preparation with SEM viewing. For this purpose, resin blocks containing the fixed sample are sectioned in an SEM specimen chamber with an ultramicrotome, and the resultant block face is imaged using back-scattered electrons produced from the interaction of the incident electron beam with the sample. By serially imaging the block face after each section is removed, one can obtain image stacks that cut through a sample. The amount of data generated from a sample with this technique is enormous – a single specimen of 1 mm^3 volume can generate several hundreds of terabytes of data. Nevertheless, the method is gaining ground as a means for high-resolution three-dimensional structures and has been used to image brain and other neuronal structures (Schwartz *et al.*, 2012).

Transmission electron microscopy is also used frequently to obtain three-dimensional structures of proteins and simple particles such as viruses. These analyses can be done in two ways, either by electron diffraction (see Section 3.3.6) or by single particle averaging. For the latter method, samples are visualised at high resolution (albeit not at atomic resolution), and the images of hundreds of small particles are mapped and modelled simultaneously using averaging methods to obtain a consensus structure. The three-dimensional envelope produced in this way can be used to guide the assembly of higher resolution molecular structures from individual monomers produced by X-ray and NMR techniques.

Molecular localisation

The major advance in biological light microscopy in the last 50 years has been the ability to tag specific molecules in cells, either directly with tags incorporated into the molecular structure or indirectly using specific markers, such as antisera, lectins or other affinity probes. These methodologies have been further complemented by the adoption of fluorescence techniques, which allow the deployment of fluorescent probes.

As electron microscopes do not operate within the visible spectrum, one cannot use probes of different colours but need to resort to molecular entities suitable for this imaging technique. Markers possessing a high electron density and regular defined shapes are thus required for use in electron microscopy and are available in the form of colloidal gold spheres. These probes are easily distinguished from native biological samples in electron micrographs due to their high atomic number Z (strong signal) and regular shape (as opposed to shapes of biological structures). The gold spheres have a net negative charge on their surface and will bind to a variety of ligands, so that particles can be made with different surface functionalisation. For example, gold particles have been conjugated to immunoglobulin or protein A for immunocytochemistry, streptavidin for *in situ* hybridisation and lectins for detecting carbohydrates. There have also been gold particles developed that incorporate a fluorescent probe, as well as a surface functionalisation molecule, so the specimen can be used for fluorescence and ultrastructural colocalisation studies.

The techniques for molecular localisation in TEM are similar to those of light microscopy except that one must keep in mind the high magnification and the nature of sample preparation. Standard TEM fixatives and

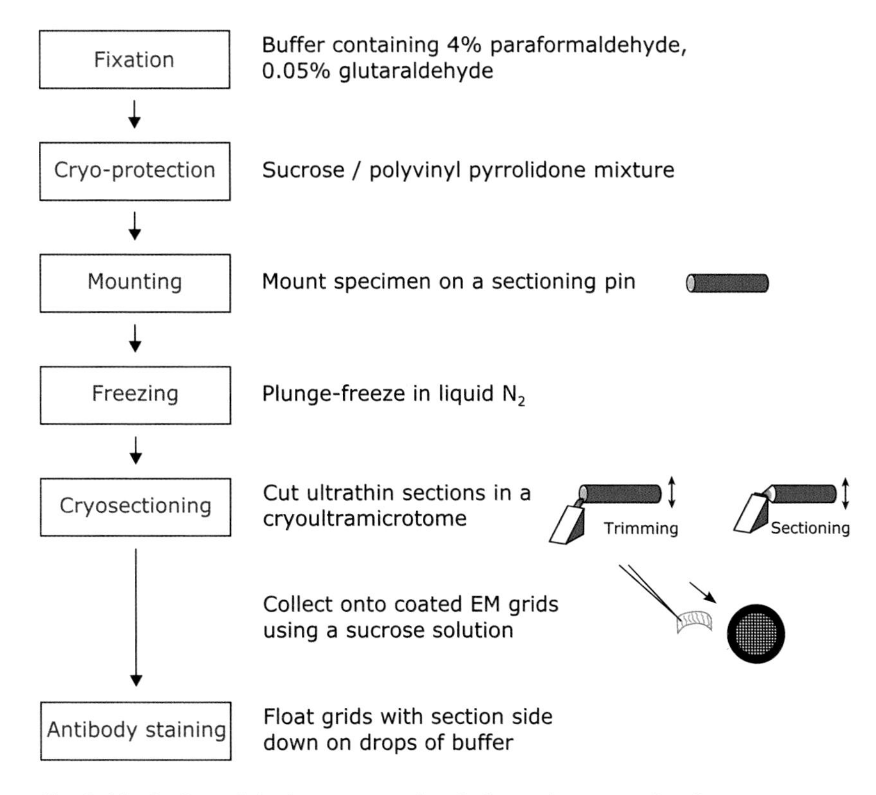

Fig. 3.14. Outline of the immunocytochemical sample preparation for immunoelectron microscopy using TEM.

resins tend to react with available binding sites on the target molecules of interest. For immunocytochemistry, where specific antibodies are used to localise proteins in a sample, preferred fixatives include paraformaldehyde, while commonly used resins include acrylate-based resins such as Lowicryl®. As fixation may potentially damage or change the structure of the sample (for example, contents may be removed during the repeated washing procedures), an alternative method of sample preparation employs a form of cryo-sectioning of cryo-protected samples. A protocol for immunocytochemistry of cryo-protected material is shown in Fig. 3.14.

3.3.5 Energy-dispersive spectroscopy

All elements – but specifically heavy elements – emit characteristic X-rays when their electrons make transitions between the lower atomic energy levels (see atomic fluorescence in Section 2.1.1). These transitions occur after bombardment of a sample with an electron beam. Therefore, many electron microscopes (transmission or scanning electron) are often also equipped with detectors to monitor the characteristic X-rays emerging from a sample. When plotting the number of X-ray photons received over time, abundant elements can be identified by having high peaks. Frequently, characteristic X-ray peaks overlap, in which case particular care must be taken to interpret the plots from certain complex tissues. These analyses can be qualitative or quantitative, the latter being more complex and laborious to perform.

Other methods for analysing and mapping elements with large atomic numbers Z in samples include EELS and high-angle annular dark-field (HAADF)-STEM. These methods filter specific electrons with particular energies from the transmitted beam by TEM or STEM. With the technique of EELS, it is possible to analyse the inelastically scattered electrons emerging from a sample. The amount of energy lost by these electrons is dependent upon the atomic number Z of elements in the sample. By specifically measuring the energy of these electrons and knowing other features of the sample, such as its thickness, it is possible to use the electron energies to map the distribution of elements in the specimen.

While elemental analysis of biological samples is a potentially valuable method, allowing high-resolution mapping and localisation of elements in biological tissues, much care needs to be taken, both in sample preparation and in subsequent analysis:

- Specimen preparation is important because many heavy elements are present in the sample only in trace amounts. These elements may be mobile in the sample, and will relocate, or simply being washed out, during specimen preparation. For this reason, many electron microscopy studies require the use of cryo-techniques to minimise sample alterations. Cryo-electron microscopy enhances the technique of electron microscopy to obtain images of objects in vitrified ice. There are multiple ways of freezing samples, including the use of a cryogen such as liquid ethane, slam freezing (where an object is slammed against a cold metal block) or high-pressure freezing (pressures of around 2000 bar suppress ice-crystal formation and permit freezing of larger samples than ambient pressure freezing techniques). The cooling process is so fast that the ice obtained does not have time to adopt a hexagonal or cubic structure but rather exhibits an amorphous, glass-like structure as with flash-frozen crystals for X-ray work. For protein molecule or virus preparations, this yields small ice films of about 100 Å thickness. In the case of biological samples, the specimen can be frozen under almost physiological conditions.
- Analyses of spectra that identify the presence of elements are prone to false results, and one needs to apply great care in analysis (for example,

due to peak overlap as mentioned above). In many cases, the methodologies require the use of several controls and standard samples for spectra comparison. These caveats aside, elemental analysis has proven very useful in electron microscopy in the life sciences. Microanalysis has been used in a wide variety of biological applications, from analysis of metal-ion content in bacterial cells to transport of metals in plant cells or marine animals. In human physiology, fundamental insight has been gained by these methods into cell physiology, particularly in areas of epithelial transport ions and the sodium/potassium balance, in toxicology of environmental pollutants and in elucidating ion balance during apoptosis.

3.3.6 Electron crystallography

This powerful method can be used hand in hand with X-ray crystallography and has been important in providing structural analysis of two-dimensional crystals and samples that are not amenable to formation of large crystals. In particular, electron crystallography has emerged as an important method for analysing the structure of proteins, most notably membrane-associated proteins. As electrons show greater interaction with matter than X-rays, they can provide substantially more information from very thin samples that appear invisible to X-rays.

In modern applications of electron crystallography, the sample is loaded into a cryo-transmission electron microscope. The specimen is first placed onto a grid coated with a thin carbon layer. There are a number of ways used to prepare the sample. Water-soluble samples may be laid over a thin lipid layer or may be trapped in a thin aqueous layer in preformed holes in the grid coatings. The specimen is frozen rapidly and, after placing the grid inside the microscope, is scanned for a protein that sits, suspended in a thin layer of frozen water, in one of the holes. During this preliminary scanning, the beam intensity is typically set to produce very low illumination, as the electron beam may damage the specimen.

Electron crystallography works on the principle that atoms in a crystal are present in an ordered array and act like a diffraction grating (see also Section 3.4.1). Relative to the wavelength of the electron beam, the atoms in the sample are spaced far apart. Most electrons will pass through the sample unscattered, but some of them will be scattered (diffracted) by the thin crystal. Within the electron microscope, in the objective lens below the sample is an aperture, called a special area aperture. It consists of a thin metal strip that has a number of holes of different sizes. By placing this strip into the beam path of the transmission electron microscope, a region of the crystal can be selected for analysis. The special aperture is placed in such a way that it blocks the entire electron beam except for one point. The microscope can then be tuned such that the image is focused into the back focal plane (see Fig. 3.11). The resulting image is the diffraction pattern of the selected crystal, with spots indicating the position of individual diffracted beams (similar to the single-crystal diffraction image in Fig. 3.17). The subsequent analysis requires substantial image processing and Fourier transformation to obtain a two-dimensional crystal structure.

Electron crystallography has been used to provide structures for many molecules. Among the first successful two-dimensional crystal structures obtained by this method were membrane-resident proteins like bacteriorhodopsin, the aquaporins, tubular structures including nicotinic acetylcholine receptor, and cytoskeletal molecules like the α,β-tubulins.

3.4 X-ray crystallography

3.4.1 General principles of diffraction

The interaction of electromagnetic radiation with matter causes the electrons in the exposed sample to oscillate. The accelerated electrons, in turn, will emit radiation of the same frequency as the incident radiation, called the secondary waves. The superposition of waves gives rise to the phenomenon of interference. Depending on the displacement (phase difference) between two waves, their amplitudes either reinforce or cancel each other out. The maximum reinforcement is called constructive interference, while the cancelling is called destructive interference. The interference gives rise to dark and bright rings, lines or spots, depending on the geometry of the object causing the diffraction. Diffraction effects increase as the physical dimension of the diffracting object (aperture) approaches the wavelength of the radiation. When the aperture has a periodic structure, for example in a diffraction grating, repetitive layers or crystal lattices, the features generally become sharper. Bragg's law (Fig. 3.15) describes the condition that waves of a certain wavelength will constructively interfere upon partial reflection between surfaces that produce a path difference only when that path difference is equal to an integral number of wavelengths (Bragg & Bragg, 1913). From the constructive interference, i.e. diffraction spots or rings, one can determine dimensions in solid materials, provided the electromagnetic radiation has a wavelength λ commensurate with the interatomic separation.

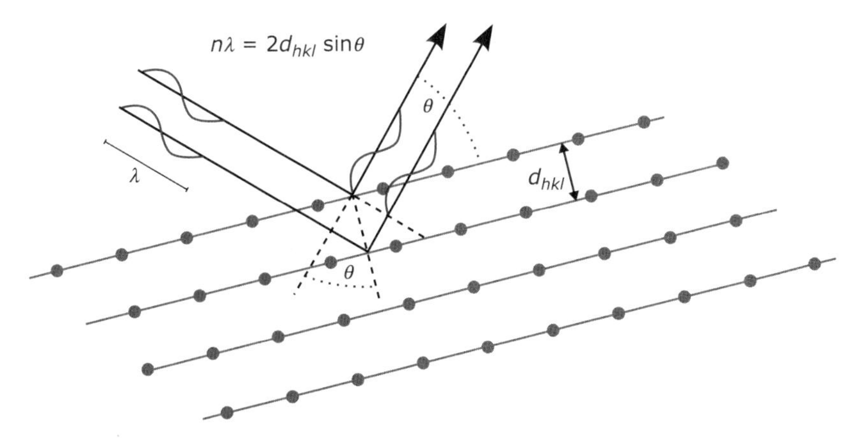

Fig. 3.15. Bragg's law (Bragg & Bragg, 1913). Interference effects are observable only when radiation interacts with physical dimensions that are approximately the same size as the wavelength of the radiation. Only diffracted beams that satisfy the Bragg condition are observable (constructive interference). Diffraction can thus be treated as selective reflection. n is an integer ('order'), λ is the wavelength of the radiation, d is the spacing between the lattice planes and θ is the angle between the incident/reflected beam and the lattice plane. h, k and l are the Miller indices that describe a particular set of diffraction planes.

As the distances between atoms or ions are in the order of 10^{-10} m (1 Å), diffraction methods used to determine structures at the atomic level require radiation in the X-ray region of the electromagnetic spectrum, or beams of electrons or neutrons with a similar wavelength. While electrons and neutrons are particles, they also possess wave properties with the wavelength depending on their energy (de Broglie hypothesis). Accordingly, diffraction can also be observed using electron and neutron beams. However, each method also has distinct features, including the penetration depth, which increases in the series electrons $<$ X-rays $<$ neutrons.

3.4.2 Fundamentals of X-ray diffraction

X-rays for chemical analysis are commonly obtained by rotating anode generators (in house) or synchrotron facilities (Fig. 3.16). In rotating anode generators, a rotating metal target is bombarded with high-energy (10–100 keV) electrons that knock out core electrons. An electron in an outer shell fills the hole in the inner shell and emits the energy difference between the two states as an X-ray photon (atomic fluorescence; see Section 2.1.1). Common targets are Cu, Mo and Cr, which have strong distinct X-ray emission at 1.54, 0.71 and 2.29 Å, respectively, which is superimposed on a continuous spectrum known as Bremsstrahlung. In synchrotrons, electrons

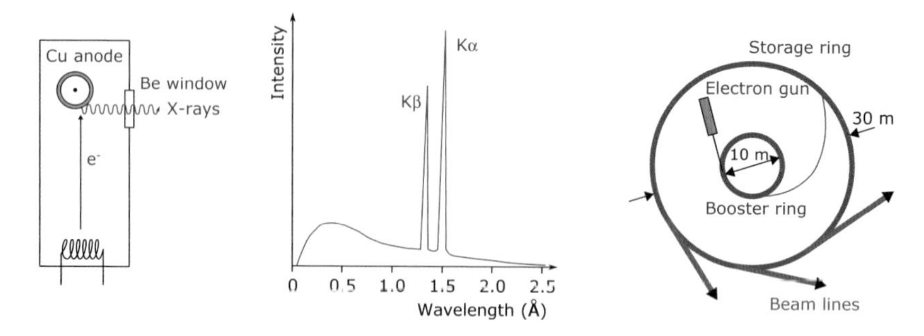

Fig. 3.16. Instrumentation for X-ray diffraction. The most common X-ray sources are rotating anode tubes (left) and particle storage rings that produce synchrotron radiation (right). The X-ray spectrum generated by a rotating Cu anode is shown in the middle. The characteristic Kα and Kβ radiation emission lines due to atomic fluorescence spike out of the underlying Bremsstrahlung spectrum.

are accelerated in a ring, thus producing a continuous spectrum of X-rays. Monochromators are required to select a single wavelength.

As X-rays are diffracted by electrons, the analysis of X-ray diffraction datasets produces an electron density map of the crystal. Because hydrogen atoms have very little electron density, they are not usually determined experimentally in protein structures by this technique; they are, however, in most small-molecule applications.

Unfortunately, the detection of light beams is restricted to recording the intensity of the beam only. Other properties, such as polarisation, can only be determined with rather complex measurements. More seriously, the phase of the light waves is systematically lost in the measurement. This phenomenon has thus been termed the phase problem owing to the essential information contained in the phase in diffraction and microscopy experiments. The X-ray diffraction data can be used to calculate the amplitudes of the three-dimensional Fourier transform of the electron density. However, only together with the phases can an electron density be calculated in a process called Fourier synthesis.

Different methods to overcome the phase problem in X-ray crystallography have been developed, including:

- molecular replacement, where phases from a structurally similar molecule are used;
- experimental methods that require incorporation of heavy element salts (multiple isomorphous replacement, MIR);
- experimental methods where methionine has been replaced by selenomethionine in proteins (multi-wavelength anomalous diffraction, MAD);

- experimental methods using the anomalous diffraction of the intrinsic sulfur in proteins (single wavelength anomalous diffraction, SAD); and
- direct methods, where a statistical approach is used to determine phases. This approach is limited to very high-resolution datasets and is the main method for small-molecule crystals as these provide high-quality diffraction with relatively few numbers of reflections.

Through solution of the phase problem, an initial electron density map is obtained. Individual atoms need to be assigned to electron density peaks in a step called map interpretation. As initial electron density maps are either biased (as in the case of molecular replacement solutions) or crude estimates (maps from experimental phasing), the model established during structure solution needs to be refined in an iterative process. Typically, this involves manual adjustments with molecular graphics software and computational refinement that performs least-squares minimisation using suitable target functions to achieve the best possible agreement between observed (F_o) and calculated (F_c) data. F_o and F_c are called structure factors, the amplitudes of which can be obtained as the square root of the diffraction intensity I. The intensity of individual diffraction spots is accessible by measuring their 'blackness' on the recorded images. One important parameter that measures the agreement between the obtained model and the experimental data is the relative deviation of F_o and F_c expressed as the R-factor:

$$R = \frac{|\sum |F_o| - |F_c||}{|\sum |F_o||}. \tag{3.3}$$

It is a crude but widely used measure of a successful crystal structure determination. While values for small molecules generally are between 0.02 and 0.05, those for a protein crystal structure typically vary between 0.15 and 0.20.

3.4.3 Single-crystal diffraction

A crystal is a solid in which atoms or molecules are packed in a particular arrangement within the unit cell, which is repeated 'indefinitely' along three principal directions in space. Crystals can be formed by a wide variety of materials, such as salts, metals, minerals and semiconductors, as well as various inorganic, organic and biological macromolecules. The unit cell is the smallest repeating volume of the crystal, akin to identical boxes stored tightly packed together, and characterised by the three basis vectors that describe the translation of the unit cell into all three dimensions. The lengths of the basis vectors (a, b and c) and the angles between them (α, β and γ) are called the cell dimensions and characteristic for a particular crystalline form of matter. In addition to this geometric periodicity, crystals possess various elements of symmetry, giving rise to seven crystal systems that are distinguished by their shapes. Considering all combinations of possible symmetries for three-dimensional crystals gives rise to 230 different groups, called space groups.

For single-crystal diffraction, crystals are grown in the laboratory, mounted on a goniometer and exposed to X-rays produced by rotating anode generators (in house) or by a synchrotron facility. A diffraction pattern of regularly spaced spots known as reflections is recorded on a detector, most frequently image plates or CCD cameras for proteins, and movable proportional counters for small molecules.

An incident X-ray beam is diffracted by a crystal such that beams at specific angles are produced, depending on the X-ray wavelength, the crystal orientation and the structure of the crystal (i.e. unit cell). The diffraction process follows Bragg's law (Fig. 3.15), and the resulting reflections (Fig. 3.17) are an image of the 'real' three-dimensional structure in reciprocal space. The geometric transformation relating 'real' three-dimensional and reciprocal space results in a different coordinate system in reciprocal space and changes all distances to their reciprocals.

In the example shown in Fig. 3.17, many individual spots are seen on the diffraction image, each being elicited by a different set of planes (described in Fig. 3.15) from a single protein crystal. The smaller the distance of a particular set of planes that causes a diffracted beam, the larger the angle between the diffracted and the primary beam becomes, and the further out on the detector plate the diffraction spot appears. Thus, the diffraction resolution extends in concentric circles from the centre of the image where

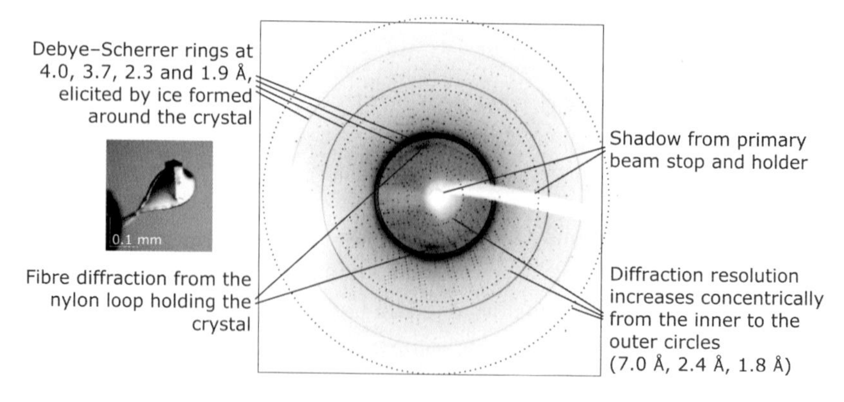

Fig. 3.17. Single-crystal diffraction image of a protein, acquired at −120 °C.

the primary beam hits. It is important to note that the resolution has nothing to do with the smallest bonding distances of atoms that can be resolved by the crystallographer. The resolution is a measure of the distance of a particular set of lattice planes that give rise to diffraction at a particular distance from the image centre. However, the higher the resolution (i.e. the smaller its value), the more fine details can be resolved in the electron density map of a crystal structure. The diffraction experiment was carried out under cryogenic conditions; therefore, the crystal was mounted in a nylon fibre loop mounted on a copper pin. The nylon fibre gives rise to some background fibre diffraction (see Section 3.4.4), which is also visible on the diffraction image. If the protein crystal has remaining water on the surface before being frozen, this can lead to solid ice formation, which diffracts X-rays like a powdered solid and thus gives rise to Debye–Scherrer diffraction rings (see Section 3.4.5). This has also been the case for the crystal used in the example shown in Fig. 3.17, where the diffraction rings elicited by ice are clearly visible.

To record a dataset, the crystal is gradually rotated and a diffraction pattern is acquired for each distinct orientation. In a process called indexing, these two-dimensional images are then analysed by identifying the appropriate reflection for each set of lattice planes (identified by its Miller indices h, k and l) and measuring its intensity. The unit cell parameters can be obtained from the separation of spot patterns, which represent the interplanar spacings of lattice planes and are characterised by their distances d_{hkl} (Fig. 3.15). Through analysis of symmetries within the spot patterns, the appropriate space group of the crystal can be determined. If information about the phases is available, these data can then be used to calculate a three-dimensional electron density within the unit cell using the mathematical method of Fourier transformation. The positions of the atomic nuclei are subsequently deduced from the electron density by computational refinement and manual intervention using molecular graphics.

3.4.4 Fibre diffraction

Certain biological macromolecules, such as for example DNA and cytoskeletal components, cannot be crystallised as single crystals but form fibres. In fibres, the axes of the long polymeric structures are parallel to each other. While this can be an intrinsic property, for example in muscle fibres, in some cases the parallel alignment needs to be induced. As fibres show helical symmetry, it is possible to analyse the diffraction from oriented fibres and deduce the helical symmetry of the molecule – and in favourable cases the molecular structure. Generally, a model of the fibre is constructed and the expected diffraction pattern is compared with the observed diffraction. Historically, fibre diffraction was of central significance through enabling the determination of the three-dimensional structure of DNA by Crick, Franklin, Watson and Wilkins.

Two classes of fibre diffraction patterns can be distinguished. In crystalline fibres (e.g. the A-form of DNA), the long fibrous molecules pack to form thin microcrystals arranged randomly around a shared common axis. The resulting diffraction pattern is equivalent to taking a long crystal and spinning it about its axis during the X-ray exposure. All Bragg reflections are recorded at once. In non-crystalline fibres (e.g. the B-form of DNA), the molecules are arranged parallel to each other but in a random orientation around the common axis. The reflections in the diffraction pattern are now a result of the periodic repeat of the fibrous molecule. The diffraction intensity can be calculated via Fourier–Bessel transformation replacing the Fourier transformation used in single-crystal diffraction.

3.4.5 Powder diffraction

Powder diffraction is a rapid method to analyse multi-component mixtures without the need for extensive sample preparation. Instead of using single crystals, the solid material is analysed in the form of a powder where, ideally, all possible crystalline orientations are equally represented. Different lattice planes, characterised by their Miller indices h, k and l, give rise to diffracted beams that emerge from the sample in the centre of the camera as a cone (Fig. 3.18). The diffraction cones are intercepted by a film surrounding the specimen in a circular fashion or by a detector moving along an equatorial arc, thus giving rise to arcs, which are sections of the so-called Debye–Scherrer rings.

With a Bragg diffraction of angle θ, every diffracted beam makes an angle of 2θ to the diffracted beam; the angle of the diffraction cone is

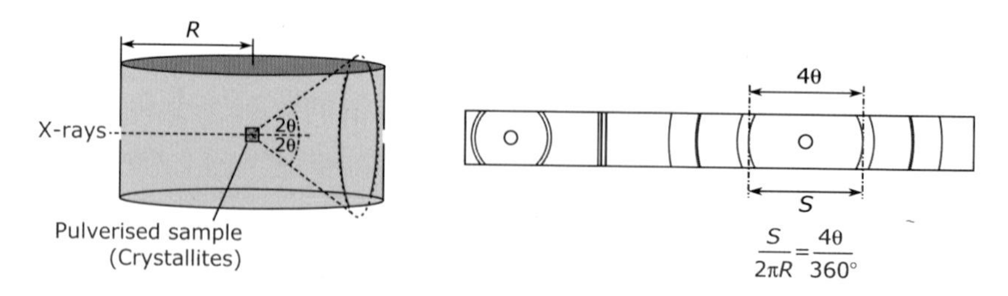

Fig. 3.18. Schematics of powder X-ray diffraction. Left: Experimental setup in a Debye–Scherrer or Guinier camera. Right: Schematic powder diagram from a Debye–Scherrer film.

thus 4θ. Every cone is detected on the film by two segments, which are positioned symmetrically around the exit and entry holes of the primary beam on the film. The higher the degree of homogenisation of the powder, the more continuous the appearance of the segments. Less homogeneous samples may only show parts of the segments, as only a few crystallites are present in a particular orientation.

The Debye–Scherrer camera used to be one of the most common instruments for obtaining powder diffraction data from a sample. The resolution of bands obtained from the Debye–Scherrer camera was considerably improved when a focusing monochromator made of curved silicon was introduced. Importantly, the X-ray beam was focused onto the film, not onto the sample. Over time, the original Debye–Scherrer camera was replaced by the Guinier camera, which used a convergent and intense primary X-ray beam requiring exposure times of several hours and subsequent film development. In many Guinier cameras, the wet-film technique dating back about 100 years has step by step been replaced by counting scintillation and proportional detectors, mainly to enable digital data acquisition. Recently, image plate detection methods have also been introduced for powder cameras, thus reducing measurement times to a few minutes.

From the camera geometry, values of the angle θ can be calculated for individual Debye–Scherrer rings with assigned Miller indices. Using appropriate equations, unit cell parameters can be determined from the measured glance angle θ and the lattice plane distances d can be assigned using the Bragg equation. Using the appropriate formulas describing the lattice plane distance d in a given crystal system, in some cases the Miller indices h, k and l can be assigned. However, in many cases, detailed indexing is not actually required. Using the determined lattice distances d and the intensity of the Debye–Scherrer rings, these data can be compared to a known standard or used to search a database (Powder Diffraction File by the International Centre for Diffraction Data, formerly known as ASTM database by the American Society for Testing and Materials) for identification of the individual components.

3.4.6 Repositories of structural data

Three-dimensional structures of molecules obtained by X-ray diffraction or NMR methods (see Section 3.2) are stored in public repositories. Deposition of these models and/or data is typically a requirement for publication of studies in the scientific literature. Different databases cover different structural techniques, as shown in Table 3.3.

Table 3.3. Databases archiving three-dimensional structures of molecules.

Database	URL	Content	Licence
Cambridge Structural Database (CSD)	http://www.ccdc.cam.ac.uk/products/csd/	Organic small-molecule crystal structures	Licence required
Inorganic Crystal Structure Database	http://www.fiz-karlsruhe.de/icsd_content.html	Inorganic crystal structures	Licence required
CRYSTMET®	http://www.tothcanada.com/	Metals and alloys	Licence required
Protein Data Bank (PDB)	http://www.rcsb.org/pdb/	Polypeptides and polysaccharides with more than 24 units from crystallographic and NMR experiments	Public domain
Biological Magnetic Resonance Data Bank (BMRB)	http://www.bmrb.wisc.edu/	Structures determined by NMR methods	Public domain
Nucleic Acids Data Bank	http://ndbserver.rutgers.edu/	Oligonucleotides	Public domain
PDF Databases	http://www.icdd.com/	Powder diffraction data	Licence required

Table 3.4. Variations of the CIF data format for particular disciplines.

Discipline	File format	Comment
Macromolecular crystallography	mmCIF	An alternative to the PDB format
Small-molecule (core) crystallography	cifdic.C91	CIF dictionary for crystallographic data
Powder diffraction	cif_pow_core.dic	CIF dictionary for powder diffraction data
Nuclear magnetic resonance	NMRif	The equivalent of CIF for NMR data
General molecular information	MIF	Complements the CIF with items describing two-dimensional structure

The most important data formats for archiving and exchanging structural information about molecules are the crystallographic information file (CIF) and Protein Data Bank (PDB) formats. These data formats can be read by molecular graphics software programs,which are used to visualise the three-dimensional structure of molecules. The CIF and PDB formats follow dictionaries that ensure correct syntax and formatting of the data presented; all these files are text-based (ASCII files). Based on the CIF format, more specialised formats have been developed for particular disciplines, described in Table 3.4.

3.5 X-ray single-molecule diffraction and imaging

X-ray crystallography has been a major workhorse for the understanding of structure and function of solid state materials as well as biological macromolecules. For example, as of September 2012, the structures of biological macromolecules deposited in the PDB exceeded 84 500, with 88% determined by X-ray crystallography and 11% determined by NMR. Crystallography requires the presence of a crystal (Section 3.4) in which the same molecule is present in an ordered pattern many millions of times. The regular pattern (the crystal lattice) serves three purposes. Firstly, the symmetry rules of three-dimensional crystallography are of fundamental importance to the techniques that enable deduction of three-dimensional structures from the observed diffraction patterns. Secondly, the lattice ensures that the X-ray dose and the accompanying potential damage is distributed over the many copies of the molecule that constitute the crystal. Thirdly, the lattice also acts like an amplifier for the diffracted beams obtained from an incident X-ray beam, which can thus be observed with appropriate detection devices. At a wavelength of 1 Å, only 10% of incident photons are actually scattered/diffracted by a sample and are thus useful for structural analysis, while 90% of photons account for radiation damage by just transferring energy onto the sample.

In 2000, the group of Hajdu suggested that diffraction from single molecules should be detectable if one uses an X-ray source of sufficient brightness ('ultrabright') and with a sufficiently short pulse (Neutze *et al.*, 2000). They proposed that novel fourth-generation X-ray sources (X-ray-free electron lasers, see Section 3.5.1) might be able to provide the required specifications.

Theoretically, it is possible to obtain high-quality diffraction images with very small crystals using synchrotron X-ray radiation (see Section 3.4.2); however, in many cases, the required high X-ray doses cause significant damage to the small crystals before a diffraction image can even be recorded. The new X-ray sources will thus need to deliver ultra-high doses (billions of times brighter than the light available from current synchrotrons) – and they will need to do that in very short pulses (duration in the order of 10–100 fs), so that diffraction data can be collected before damage occurs in the sample.

With the first fourth-generation light source commencing work in 2009, initial work had focused on the testing of theoretical predictions. It soon became clear that indeed ultrashort X-ray pulses lead to measurable diffraction of single objects (nanostructures, cells and viruses) that allow reconstruction of phases and thus determination of structures. The ultrahigh X-ray doses delivered lead to vaporisation and thus total destruction of the single objects but not before the elastically scattered X-rays escape the point of impact.

The first proof of principle was obtained when microcrystals of photosystem I, an integral membrane protein complex that captures light to mediate electron transfer from plastocyanin to ferredoxin in plants, were subjected to serial femtosecond crystallography (SFX) at the Linac Coherent Light Source (LCLS) in California, USA, and interpretable electron density maps were obtained (Chapman *et al.*, 2011). However, as the X-ray light used in those early experiments had a rather long wavelength of 6.9 Å, the resolution of the collected data was restricted to about 8 Å – a low-resolution structure that does not reveal finer atomic details. With the very recent development of a novel imaging facility (the LCLS coherent X-ray imaging instrument) that provides hard X-ray pulses, wavelengths of up to 1.32 Å (cf. Cu-Kα radiation: 1.54 Å) became possible. The first protein to be successfully tested with this new technology was lysozome, and its structure was determined by SFX at a resolution of 1.9 Å using microcrystals passing through the beam within a liquid jet in random orientation (Boutet *et al.*, 2012). When compared with known three-dimensional structures of lysozyme, that determined by SFX showed no signs of radiation damage, proving that single-molecule diffraction can be used to determine three-dimensional structures of molecules at an atomic level.

3.5.1 X-ray-free electron laser

Imaging techniques such as X-ray crystallography rely on light waves of the same wavelength that are exactly in phase (coherent light), and the resolution of the image obtained is a function of the exposure to light energy (dose), with radiation damage setting the maximum dose that can be used. The light available at contemporary synchrotrons is not able to deliver the coherent intensities required for imaging with a single pulse of light. Linear electron accelerators provide an alternative approach for X-ray generation. These instruments were conceived in the 1920s and have been used widely in particle physics and medical applications; they are typically the source of electrons injected into synchrotron rings. As electrons in a linear accelerator do not recirculate and thus gradually lose energy, very bright electron beams can be generated. Because brightness is determined by the physical components of spot size, energy spread and pulse duration, it is conceivable that shorter pulses will lead to brighter beams. With X-ray-free electron lasers, a femtosecond pulse of high-energy electrons ('electron bunch') is used as the lasing medium. The electrons are accelerated in an oscillating magnetic field, which causes them to spontaneously emit X-ray photons. Through interaction of the emitted X-ray light with the electron bunch, many microbunches are formed that all emit X-rays in a coherent fashion. The pulses obtained from the microbunches (at the order of picoseconds) thus combine to one macrobunch of coherent light obtained as a pulse at 10–100 fs. Intriguingly, as the microbunch pulses arise at the picosecond scale, there is room for further technical development to obtain subfemtosecond X-ray pulses and thus image atomic dynamics on an unprecedented timescale.

Current SFX experiments are conducted with 120 pulses s^{-1} (120 Hz) and each pulse delivers a dose of about 50 MGy onto a single diffracting object. This dose corresponds to a surface power density of 1.2×10^{17} W cm^{-2}, which is in the order of magnitude of the total power of the Earth receiving by the sun (2×10^{17} W), concentrated into 1 cm^2. For comparison, a single crystal in protein crystallography receives a dose in the order of 25 MGy over an experiment with 20 min total exposure time. In terms of total energy deposited on the diffracting object, SFX and conventional crystallography are thus comparable.

3.5.2 Sample handling

Single-molecule diffraction and imaging techniques harbour three major challenges when it comes to sample handling:

- As everything in the beam will be imaged, and contributions from any non-sample object interfere with the reconstruction of diffraction patterns and phases, samples need to be delivered into the beam without containers or artificial frameworks.
- Single objects need to be delivered into the beam, and therefore appropriate delivery formats need to be developed.
- Objects delivered into the beam rotate and move freely in space, thus making it impossible to know their orientation and position.

Much knowledge was gained from techniques with similar requirements, such as mass spectrometry and electron microscopy. In electron cryo-microscopy, studies on hydrated samples in high vacuum at low temperatures are performed. Mass spectrometry already uses electrospray and nano-flow techniques to deliver samples into the instruments. Current SFX experiments with protein nanocrystals use a liquid jet of about 4 μm diameter that delivers the sample into the X-ray beam (~3 μm diameter). Notably, measurements are done at room temperature, so there is no lengthy sample preparation required or preservation required. Present experiments with laser pulse rates of 120 Hz use in the order of 1.5 mg protein in the form of nanocrystals to deliver data acquired over a period of 15 min. It is envisioned that future facilities such as the European FEL in Hamburg will be able to deliver the same amount of data from about 50 μg of protein in 30 s.

At the current stage, the orientation/position problem seems to be overcome without experimental control methods. The positioning of exactly one object in the X-ray beam may be possible by the fact that the spot size of the X-ray beam is extremely small and the pulse duration extremely short, as well as a fine-tuning of the delivery jet speed and the sample dilution. Proof-of-principle experiments with nanocrystal diffraction by SFX show that it is possible to solve the orientation problem by purely computational means.

3.5.3 Diffraction detection and analysis

The main challenges of diffraction data recording from X-ray sources with ultrahigh intensity stem from the destructive power of the X-ray pulses as well as from the extremely high repetition rates of exposure. Recently developed detector types are composed of multiple small single-photon CCD detectors arranged in a fixed or movable way around the propagation direction of the primary beam (Fig. 3.19).

Many tens of thousands of diffraction snapshots are required to obtain a full set of diffraction data, thus coining the term serial crystallography. As the individual microcrystals brought into the beam are hit by very short light pulses and destroyed in the process, there is no time to emulate the crystal oscillation typically done in conventional crystallography. Additionally, as the individual microcrystals can adopt any orientation with respect to the incident beam, there is no correlation between the diffraction patterns

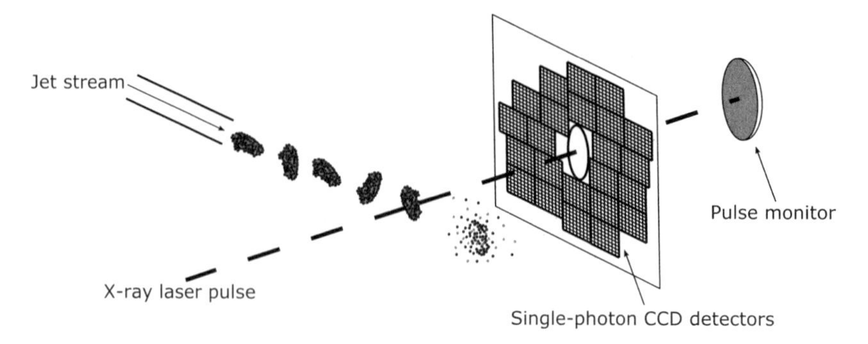

Fig. 3.19. Schematics of an X-ray-free electron laser diffraction experiment.

obtained from different crystals. When overlaying all diffraction patterns obtained from an experiment, the resulting image is essentially that obtained in powder diffraction, i.e. characteristic Debye–Scherrer rings are visible. Therefore, the final diffraction pattern for a sample needs to be reconstructed from many snapshot patterns whose orientation with respect to each other needs to be resolved and which contain only partial intensities. For the mathematical and statistical treatment of data, there are also several aspects to be considered that arise from the extremely small sample size. A first attempt at serving the needs of SFX has resulted in software packages such as CrystFEL (White *et al.*, 2012), which interfaces with conventional crystallography procedures and software programs at a certain point. Computing speed is certainly another issue as very complex calculations need to be performed on millions of individual diffracted beams.

3.5.4 Future developments

Single-molecule diffraction and serial crystallography are emerging techniques with a wide field of applications, including imaging of single cells or parts thereof and three-dimensional structures of proteins that are difficult to crystallise, as well as time-resolved structures that will allow unprecedented insights into ongoing chemical changes within a molecule. The future developments will thus be two-fold: advancing the methodology to a stage where single molecules (instead of micro- or nanocrystals) can be imaged, and technological advances that will bring about ultrafast detection devices and more convenient X-ray sources.

Contemporary in-house X-ray generators such as rotating anode machines (see Section 3.4.2) or medical X-ray devices produce continuous, rather weak light. Therefore, SFX experiments are currently restricted to linear accelerator facilities where X-ray-free electron lasers are currently being developed. However, two types of technology developed over the past decade may one day enable femtosecond serial crystallography in laboratory environments. One of these methods uses an ionised atomic plasma that can generate highly monochromatic directed laser beams, albeit current wavelengths are restricted to about 100–500 Å (cf. Cu-Kα radiation: 1.54 Å). The other method uses an intense femtosecond laser as a primer to excite electrons in a gaseous medium in a highly non-linear fashion. As a result, high harmonics of the priming laser light are emitted as a coherent, laser-like beam. This technology can already deliver wavelengths spanning the region from UV to X-rays (Popmintchev *et al.*, 2012). One may expect that these technologies will enable femtosecond imaging and diffraction in home laboratories in the next 10–20 years.

3.6 Small-angle scattering

While crystals present ordered arrays that allow high-resolution transforms and thus the detection of individual diffraction peaks (Bragg peaks), the situation is different with single molecules oriented in a random fashion such as in solution. The scattering of light off molecules without a predominant orientation yields a spherically averaged molecular transform, as observed, for example, with the diffuse patterns from X-ray scattering from biological molecules or nanostructures. An advantage for protein structure determination is the fact that samples in aqueous solution can be assessed.

In Section 2.7, we learned that incident light scattered by a particle in the form of Rayleigh scattering has the same frequency as the incident light. It is thus called elastic light scattering. The light-scattering techniques discussed in Section 2.7 have used a combination of visible light and molecules, so that the dimension of the particle is smaller than the wavelength of the light. When using light of smaller wavelengths such as X-rays, the overall dimension of a molecule is large compared with the incident light. Electrons in the different parts of the molecule are now excited by the incident beam with different phases. The coherent waves of the scattered light therefore show an interference that is dependent on the geometrical shape of the molecule. As a result:

- In the forward direction (at $\theta = 0°$), there is no phase difference between the waves of the scattered light, and one observes maximum positive interference, i.e. highest scattering intensity.
- At small angles, there is a small but significant phase difference between the scattered waves, which results in diminished scattering intensity due to destructive interference.

Small-angle X-ray and neutron scattering (SAXS or SANS) are experimental techniques used to derive size and shape parameters of large molecules. The scattering of a particle can be viewed as the interference pattern produced by all light waves or neutrons detected as coming from the particle. This pattern oscillates in a manner that is characteristic for the shape of the particle and is thus called the particle's form factor (Fig. 3.20). Both X-ray and neutron scattering are based on the same physical phenomenon, i.e. scattering due to differences in scattering mass density between the solute and the solvent ('contrast') or indeed between different molecular constituents. The extent to which X-rays or neutrons are scattered by atoms is called scattering length and depends on the specific properties of individual atoms. As X-rays are scattered by electrons in the atomic orbitals, the X-ray scattering length of atoms is proportional to the number of electrons. Neutrons, in contrast, are scattered by the atomic nuclei, and the scattering length of individual elements, and indeed isotopes, varies in a non-correlated fashion. In order to assess three-dimensional objects in solutions, one defines the scattering length density as the sum of the scattering lengths over all atoms within a given volume, divided by that volume. The contrast can then be calculated as the squared difference in scattering length density between particle and solvent.

Experimentally, a monodisperse solution of macromolecules is exposed to either X-rays (wavelength $\lambda \approx 0.15$ nm) or thermal neutrons ($\lambda \approx 0.5$ nm). The intensity of the scattered light is recorded as a function of momentum transfer:

$$q = 4\pi \frac{\sin\theta}{\lambda}, \quad (3.4)$$

where 2θ is the angle between the incident and scattered radiation (see Fig. 3.20). Due to the random positions and orientations of the particles, an isotropic intensity distribution is observed that is proportional to the scattering from a single particle averaged over all orientations. As the neutron scattering lengths of the proton (^{1}H) and the deuteron (^{2}H = D) are markedly different, the contrast in neutron scattering experiments can be varied using H_2O/D_2O mixtures or selective deuteration to yield additional information. At small angles, the scattering curve is a rapidly decaying function of q and is essentially determined by the particle shape. Fourier transformation of the scattering function yields the so-called pair distance distribution function (or size distribution function), which is a

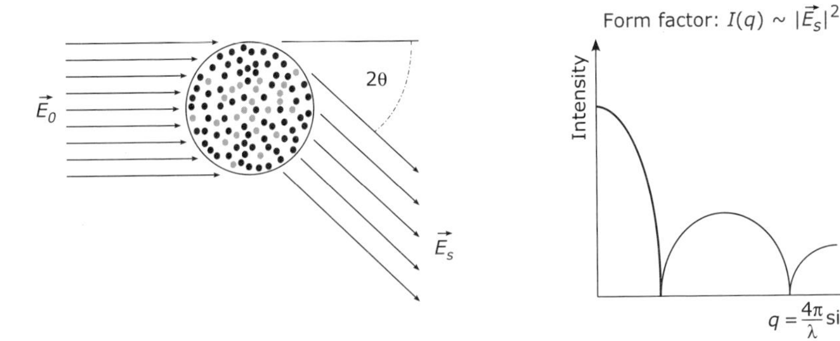

Fig. 3.20. The scattering of a particle, consisting of many atoms, can be explained as the interference pattern produced by all waves arising from every atom inside the particle. The squared sum of all wave amplitudes E_s yields an oscillating function that is characteristic for the shape (form) of the particle: the form factor. To match the experimentally determined intensity, the form factor needs to be scaled with a constant.

histogram of interatomic distances. Comparison of the pair distance distribution function with the particle form factor (the interference pattern of light coming from the particle) for regular geometrical bodies allows conclusions as to the shape of the scattering particle. Through analysis of the scattering function, one can further determine the radius of gyration R_g (average distance of the atoms from the centre of gravity of the molecule) and the mass of the scattering particle from the scattering in the forward direction.

3.6.1 Shape restoration

Computer programs have been developed that enable the calculation of three-dimensional structures from the one-dimensional scattering data obtained by small-angle scattering. Due to the low resolution of small-angle scattering data, the structural information is restricted to the shape of the scattering molecules. Furthermore, the scattering data do not imply a single, unique solution. The reconstruction of three-dimensional structures might thus result in a number of different models. One approach is to align and average a set of independently reconstructed models, thus obtaining a model that retains the most persistent features and, where possible, compare with structures obtained by electron microscopy or X-ray crystallography.

FURTHER READING

Electron spin resonance

Matsumoto, K., Subramanian, S., Murugesan, R., Mitchell, J. B. & Krishna, M. C. (2007). Spatially resolved biologic information from *in vivo* EPRI, OMRI, and MRI. *Antioxidants & Redox Signaling* 9, 1125–41.

Schiemann, O. & Prisner, T. F. (2007). Long-range distance determinations in biomacromolecules by EPR spectroscopy. *Quarterly Reviews of Biophysics* 40, 1–53.

Zhou, Y. T., Yin, J. J. & Lo, Y. M. (2011). Application of ESR spin label oximetry in food science. *Magnetic Resonance in Chemistry* 49, S105–12.

Websites on electron spin resonance

Electron Spin Resonance: http://hyperphysics.phy-astr.gsu.edu/hbase/molecule/esr.html

Electron Paramagnetic Resonance/Electron Spin Resonance (EPR/ESR): http://www.niehs.nih.gov/research/resources/epresr/index.cfm

Nuclear magnetic resonance

Blamire, A. M. (2008) The technology of MRI – the next 10 years? *British Journal of Radiology* **81**, 601–17.

Ishima, R. & Torchia, D. A. (2000) Protein dynamics from NMR. *Nature Structural Biology* **7**, 740–3.

McDermott, A. & Polenova, T. (2007) Solid state NMR: new tools for insight into enzyme function. *Current Opinion in Structural Biology* **17**, 617–22.

Pretsch, E., Bühlmann, P. & Badertscher, M. (2009). *Structure Determination of Organic Compounds*, 4th edn. Berlin, Heidelberg: Springer Verlag. (A compendium with tables of spectral data.)

Skinner, A. L. & Laurence, J. S. (2008). High-field solution NMR spectroscopy as a tool for assessing protein interactions with small molecule ligands. *Journal of Pharmaceutical Sciences* **97**, 4670–95.

Spiess, H. W. (2008). NMR spectroscopy: pushing the limits of sensitivity. *Angewandte Chemie International Edition* **47**, 639–42.

Websites on nuclear magnetic resonance

Organic Chemistry at CU Boulder: The Theory of NMR: http://orgchem.colorado.edu/Spectroscopy/Spectroscopy.html

Organic Chemistry On Line: http://www.cem.msu.edu/~reusch/VirtualText/Spectrpy/nmr/nmr1.htm#nmr1

The Basics of NMR: http://www.cis.rit.edu/htbooks/nmr/

NMR CTP Tutorial Home: http://arrhenius.rider.edu/nmr/NMR_tutor/pages/nmr_tutor_home.html

NMR Spectroscopy Theory: http://teaching.shu.ac.uk/hwb/chemistry/tutorials/molspec/nmr1.htm

Electron microscopy

Allen, T. D. (ed.) (2008). *Introduction to Electron Microscopy for Biologists*. Methods in Cell Biology, **Vol. 88**. San Diego, CA: Academic Press.

Goldstein, J., Newbury, D. E., Joy, D. C. *et al.* (2003). *Scanning Electron Microscopy and X-ray Microanalysis*, 3rd edn. Berlin: Springer Verlag.

Griffiths, G. (1993). *Fine Structure Immunocytochemistry*. Berlin: Springer Verlag.

Kou, J. (2007). *Electron Microscopy – Methods and Protocols*. Methods in Molecular Biology, **Vol. 369**, 2nd edn. Totowa: Humana Press.

McIntosh, J. R. (2007). *Cellular Electron Microscopy*. Methods in Cell Biology, **Vol. 79**. San Diego, CA: Academic Press.

Nannenga, B. L., Iadanza, M. G., Vollmar, B. S. & Gonen, T. (2013). Overview of electron crystallography of membrane proteins: crystallization and screening strategies using negative stain electron microscopy. In *Current Protocols in Protein Science*, Chapter 17, Unit 17.15.

Williams, D. B. & Carter, C. B. (2009). *Transmission Electron Microscopy: A Textbook for Materials Science*. Berlin: Springer Verlag.

Websites on electron microscopy

Transmission electron microscopy: http://www.matter.org.uk/tem/

Tomographic Reconstruction of SPECT Data: filtered back projection, algebraic reconstruction methods: http://www.owlnet.rice.edu/~elec539/Projects97/cult/report.html

Filtered back projection in SPECT: http://www.youtube.com/watch?v=MTBhqcVjQ8Q

Central Microscopy Research Facility (The University of Iowa) – Methodology: http://cmrf.research.uiowa.edu/methodology

X-ray crystallography

Massa, W. (2004). *Crystal Structure Determination*, 2nd edn. Berlin, Heidelberg: Springer Verlag.

Rhodes, G. (2006). *Crystallography Made Crystal Clear – A Guide for Users of Macromolecular Models*, 3rd edn. Amsterdam, Boston, Heidelberg: Associated Press.

Rupp, B. (2009). *Biomolecular Crystallography: Principles, Practice and Application to Structural Biology*. Abingdon, New York: Garland Science, Routledge.

Wlodawer, A., Minor, W., Dauter, Z. & Jaskolski, M. (2008). Protein crystallography for non-crystallographers, or how to get the best (but not more) from published macromolecular structures. *FEBS Journal* **275**, 1–21.

Websites on X-ray crystallography

X-rays: http://www.colorado.edu/physics/2000/xray/index.html

X-ray diffraction: http://www.matter.org.uk/diffraction/x-ray/default.htm

Powder Diffraction on the Web: http://pd.chem.ucl.ac.uk/pdnn/pdindex.htm#inst1

A Hypertext Book of Crystallographic Space Group Diagrams and Tables: http://img.chem.ucl.ac.uk/sgp/mainmenu.htm

The Crystallographic Information File (CIF): http://www.sdsc.edu/pb/cif/cif.html

X-ray single-molecule diffraction and imaging

Chapman, H. N., Fromme, P., Barty, A. *et al.* (2011). Femtosecond X-ray protein nanocrystallography. *Nature* **470**, 73–7.

Websites on X-ray single-molecule diffraction and imaging

X-ray-free electron lasers: principles, properties and applications: http://www-ssrl.slac.stanford.edu/stohr/xfels.pdf

The European X-ray laser project: http://www.xfel.eu/

Small-angle scattering

Fitter, J., Gutberlet, T. & Katsaras, J. (eds) (2006). *Neutron Scattering in Biology – Techniques and Applications.* Berlin, Heidelberg, New York: Springer.

Lindner, P. & Zemb, T. (2002). *Neutron, X-rays and Light. Scattering Methods Applied to Soft Condensed Matter.* Revised subedition. North-Holland Delta Series.The Netherlands: Elsevier. (In-depth coverage of theory and applications of light scattering at expert level.)

Lipfert, J. & Doniach, S. (2007). Small-angle X-ray scattering from RNA, proteins, and protein complexes. *Annual Review of Biophysics and Biomolecular Structure* **36**, 307–27.

Neylon, C. (2008). Small angle neutron and X-ray scattering in structural biology: recent examples from the literature. *European Biophysics Journal* **37**, 531–41.

Putnam, C. D., Hammel, M., Hura, G. L. & Tainer, J. A. (2007). X-ray solution scattering (SAXS) combined with crystallography and computation: defining accurate macromolecular structures, conformations and assemblies in solution. *Quarterly Reviews of Biophysics* **40**, 191–285.

Websites on small-angle scattering

SANS Tutorials and presentations: http://www.ncnr.nist.gov/programs/sans/tutorials/index.html

EMBO Practical Course on Solution Scattering from Biological Macromolecules: http://www.embl-hamburg.de/workshops/2001/EMBO/

The SAXS Guide (Anton Paar): http://www.ill.eu/nc/news-events/past-events/2012/bombannes-2012/sponsor-material-2012/?cid=38763&did=56873&sechash=bf67e1073.1

4 Physical methods

4.1 Centrifugation

4.1.1 Physics of centrifugation

With centrifugation, it is possible to separate molecules based on size, shape, density and viscosity. The effect of sedimentation – the separation of denser particles suspended in a less dense medium – is directly accessible from everyday experience due to the Earth's gravitational field ($g = 981\ \text{cm s}^{-2}$). In a centrifuge, the gravitational field $\vec{G}$ can be increased significantly beyond the value of the Earth's gravitational field, and its value depends on the angular velocity ω of the centrifuge, as well as the distance r of the particle from the axis of rotation:

$$G = \omega^2 r \tag{4.1}$$

The angular velocity can be calculated from the rotor speed s_{rotor} (measured in revolutions per time):

$$\omega = \frac{2\pi s_{\text{rotor}}}{60}. \tag{4.2}$$

Therefore,

$$G = \frac{4\pi^2 s_{\text{rotor}}^2 r}{3600}. \tag{4.3}$$

Generally, the centrifugal field $\vec{G}$ is expressed as multiples of the Earth's gravitational field, leading to the relative centrifugal field (RCF):

$$\text{RCF} = \frac{G}{g} = \frac{4\pi^2 s_{\text{rotor}}^2 r}{3600 \times 981} \tag{4.4}$$

with s in the unit of revolutions min^{-1}.

During centrifugation, a particle suspended in medium experiences a variety of forces. The centrifugal force $\vec{F}_{\text{centr}}$ accelerates the particle away from the axis of rotation. The centrifugal force at a revolution speed of ω and a distance r from the centrifuge spindle is:

$$F_{\text{centr}} = m_{\text{particle}} \omega^2 r. \tag{4.5}$$

Table 4.1. Average values of partial specific volumes for biological macromolecules.

Macromolecule	$v_{particle}$ (ml g^{-1})
Protein	0.73
Polysaccharide	0.61
RNA	0.53
DNA	0.58

The buoyant force $\vec{F}_{buoyant}$ acts in the opposite direction and arises from the fact that a submerged object displaces a volume of the medium equal to its own volume. The buoyant force equals the weight of the displaced medium and can thus be calculated as:

$$F_{buoyant} = -m_{solvent}\omega^2 r = -m_{particle}v_{particle}\rho_{solvent}\omega^2 r. \tag{4.6}$$

Alternatively, the volume of the particle can be calculated from its partial specific volume $v_{particle}$, thus relating the buoyant force to the mass of the particle.

For the most accurate results, the partial specific volume should be experimentally determined. This can be achieved by determination of the density of solutions with known concentrations. However, as these measurements typically require large amounts of the macromolecule to be investigated – which in many cases are not available – the partial specific volumes can also be calculated based on the sample composition and the partial specific volumes of the components comprising the macromolecule. Some average values for biological macromolecules are listed in Table 4.1.

Any movement of the submerged particle in the medium will also give rise to the frictional force $\vec{F}_R$, which is dependent on the velocity v and – via the friction coefficient f – also dependent on the viscosity η of the solvent (Stokes' law of friction, 1865):

$$F_R = -fv = -6\pi\eta_{solvent}r_{particle}v. \tag{4.7}$$

Note that, in this equation, $r_{particle}$ is the radius of the assumed spherical particle. The effective force experienced by the particle during sedimentation is thus:

$$\vec{F}_{effective} = \vec{F}_{centr} + \vec{F}_{buoyant} + \vec{F}_R. \tag{4.8}$$

The particle will float in the medium at a particular distance from the axis of rotation in the centrifuge tube, where $\vec{F}_{effective}$ is zero, i.e. all forces are in balance (Fig. 4.1). This state delivers the following equation:

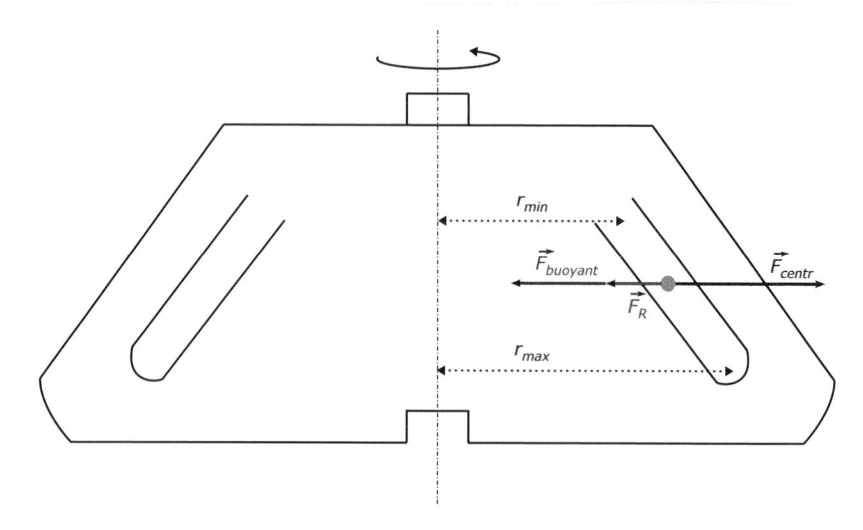

Fig. 4.1. Scheme of a fixed-angle centrifuge rotor. Shown are the radial distances $r_{\min}$ and r_{max}, as well as the effective force on a particle as a vector diagram of the three forces operating in centrifugal experiments.

$$m_{\text{particle}}\omega^2 r - m_{\text{particle}} v_{\text{particle}} \rho_{\text{solvent}} \omega^2 r - f v = 0, \tag{4.9}$$

from which we get the following after rearranging:

$$\frac{m_{\text{particle}}(1 - v_{\text{particle}}\rho_{\text{solvent}})}{f} = \frac{v}{\omega^2 r} \equiv s \tag{4.10}$$

with the unit of time, i.e. $[s] = 1$ s.

This defines the sedimentation coefficient s, which describes the velocity of the particle per unit gravitational acceleration. The sedimentation coefficient is a property of the particle and is independent of the experimental conditions. Values of s for different substances are in the range of 1×10^{-13}–100×10^{-13} s, which led to the definition of the Svedberg unit, which is defined as:

$$1S = 1 \times 10^{-13}\text{s}. \tag{4.11}$$

Serum albumin ($M = 66.5$ kDa), for example, has a sedimentation coefficient of $s = 4.5 \times 10^{-13}\,\text{s} = 4.5\,\text{S}$.

In the process of centrifugation, separated particles will start to accumulate at the outermost wall of the centrifuge tube; the location of the sediment will depend on the type of rotor being used (Fig. 4.2). As this increases the local concentration, the process of diffusion will start and counteract the sedimentation process. Over time, the two processes will be in equilibrium throughout the tube, and the concentration of the ideal single-solute component will increase exponentially with decreasing distance from the sediment.

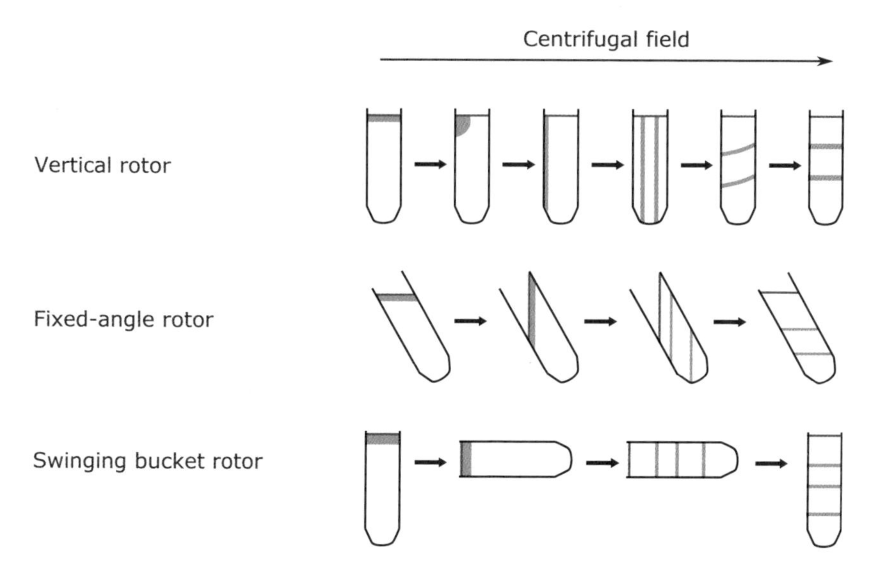

Fig. 4.2. Rotor types and sample sedimentation behaviour.

If the densities of a solute component and the solvent are equal, then the term $(1 - v_{particle}\rho_{solvent})$ becomes zero, and there will be no movement of the solute component in either direction in the gravitational field. This phenomenon is used in the technique of density-gradient sedimentation (see Section 4.1.5) to determine experimentally the density of macromolecules.

4.1.2 Analytical ultracentrifugation

Analytical ultracentrifugation allows the monitoring of sedimentation of macromolecules in a centrifugal field. The method is typically applied when hydrodynamic properties of macromolecules need to be characterised. Its particular advantage lies in the fact that macromolecules can be studied in solution without the requirement to interact with matrices or surfaces (e.g. in size-exclusion chromatography, quartz crystal microbalance), and without the need for any reference standards.

Data obtained from analytical ultracentrifugation experiments comprise a concentration distribution over the radial positions within a centrifugation tube; the data may also be collected over time. It is thus necessary to measure the concentration of the sample within the centrifugation tube *in situ*, i.e. during the centrifugation. Three types of detection method have been in use:

- Schlieren optics, a refractometric method based on the effect that light passing through areas of high refractive index (i.e. high concentration) is deviated radially;
- Rayleigh interference optics, another refractometric method that relies on the differences of the speed of light when passing areas of different refractive index; and
- absorption optical systems that allow application of the methodology discussed in Section 2.2.

The refractometric methods are of historical importance, but most modern analytical ultracentrifuges use an absorption system, which in principle allows determination of absolute concentrations at any radial position, rather than the concentration difference measurements with respect to reference points.

In terms of instrumentation, an analytical ultracentrifuge combines a UV/Vis spectrometer with a sample that is rotating at very high speed. One main aspect of these instruments is thus the mechanical design and craftsmanship, as a high-precision measurement needs to be taken in the vicinity of a massive unit rotating at very high speeds. The general layout of an analytical ultracentrifuge is shown in Fig. 4.3. A xenon (Xe) lamp

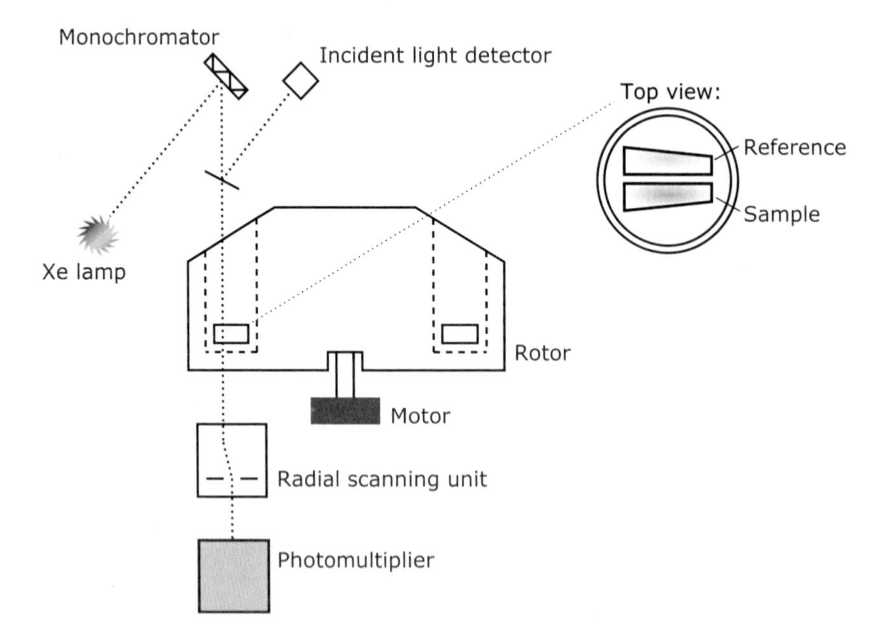

Fig. 4.3. Schematic diagram of an analytical ultracentrifuge. The light path is indicated as a dotted line. The inset shows a double sector cell with reference and sample compartments.

and a monochromator provide the light that passes through the sample and reference cells, which are called sectors. Both sectors sit next to each other in one compartment of the rotor and both are analysed in an alternating fashion. The timing information as to when one of the compartments passes through the light path is provided by magnets in the base of the rotor. The imaging system conducts a radial scan of the cell when it is in the read-out position. Alternatively, a wavelength scan can be conducted at a particular radial distance.

Double sector cells – one compartment filled with sample solution and the other with solvent – are the most popular cell types used. They allow correction against absorbing components in the solvent, as well as correction of redistribution of solvent components. The reference sector is typically filled slightly more than the sample sector in order to avoid one meniscus obscuring the other.

4.1.3 Sedimentation velocity

This type of experiment applies a sufficiently high centrifugal speed to cause rapid sedimentation of the solute towards the bottom of the cell. The concentration of solute at the meniscus will decrease and a boundary between this depleted region and the sediment will form (Fig. 4.4).

The rate of migration of the boundary, v, can be measured and allows the calculation of the sedimentation coefficient s according to:

$$s = \frac{v}{\omega^2 r}. \tag{4.12}$$

In order to remove the effects of inter-particle interactions, measurements are done at different sample concentrations, and the sedimentation coefficient is extrapolated to zero concentration, yielding s^0. Using this result, it is possible to determine the molar mass M of the sample under investigation by:

$$M = \frac{s^0 \mathrm{R} T}{D(1 - v_{\mathrm{particle}} \rho_{\mathrm{solvent}})} \tag{4.13}$$

where R is the gas constant, T the temperature and D is the diffusion coefficient given by:

$$D = \frac{\mathrm{R}T}{\mathrm{N_A} f}, \tag{4.14}$$

where $\mathrm{N_A}$ is Avogadro's constant. The diffusion coefficient is most accurately determined from light-scattering experiments (see Section 2.6.2).

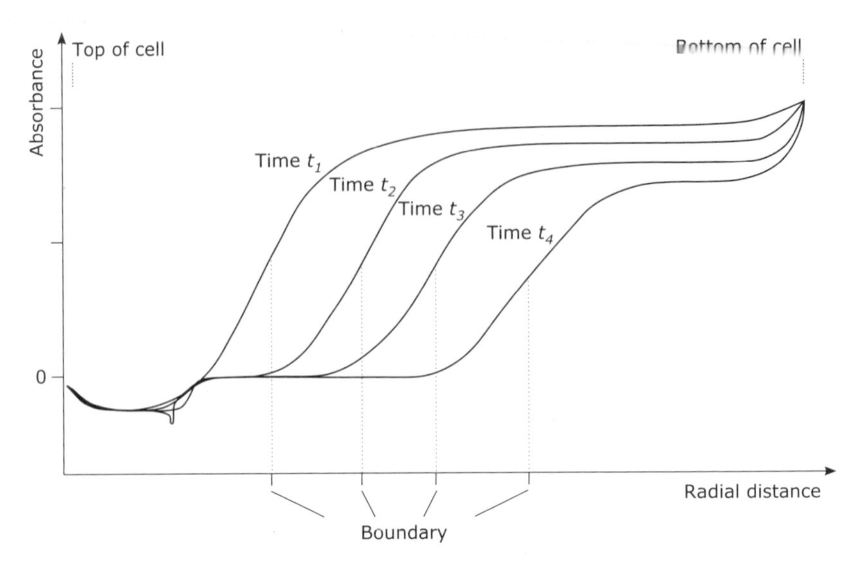

Fig. 4.4. In a sedimentation velocity experiment, the movement of the boundary (indicated by the point of inflection for each profile) can be analysed over time. As the boundary migrates down the cell, the concentration in the plateau decreases due to dilution of the sample along the radial distance. The boundary region broadens owing to the diffusion process.

4.1.4 Sedimentation equilibrium

For sedimentation equilibrium experiments, the solute is subjected to centrifugal speeds lower than required for a sedimentation velocity experiment. Under these conditions, the sedimentation process starts, but diffusion opposing the sedimentation leads to a state of equilibrium after an appropriate period of time. When in equilibrium, the concentration distribution of solute in the cell is exponential to the square of the radial position (Fig. 4.5).

The theoretical treatment of this phenomenon involves the application of thermodynamics, specifically the gradient of chemical potential, which depends on the gradient of concentration. For a non-associating ideal solute, the molar mass can be computed by:

$$M = \frac{2RT}{(1-\nu_{\text{particle}}\rho_{\text{solvent}})\omega^2} \times \frac{d(\ln\rho^*)}{dr^2}. \tag{4.15}$$

Note that in this equation the mass concentration ρ^* (at a particular radial distance r) is used instead of the molar concentration. A plot of ln ρ^* versus r^2 therefore yields a linear correlation where the slope is proportional to the molar mass M.

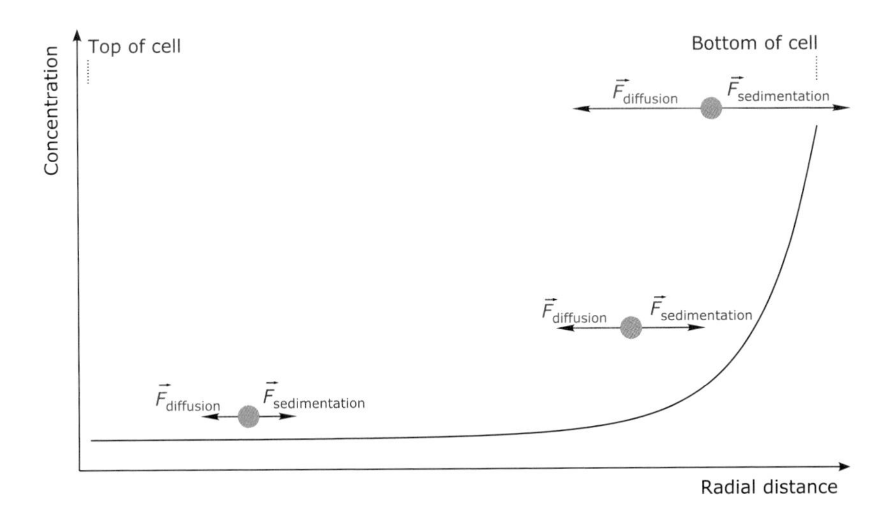

Fig. 4.5. The concentration distribution of solute in sedimentation equilibrium experiments increases exponentially with the radial distance. The forces experienced by solute molecules increase with increasing radial distance. The sedimentation and diffusion forces balance each other in equilibrium.

4.1.5 Density-gradient sedimentation equilibrium

As mentioned above, if the densities of the solute and the solvent are equal, the term $(1 - v_{particle}\rho_{solvent})$ becomes zero, and solutes will float. The method exploiting this state is called density-gradient sedimentation equilibrium or isopycnic sedimentation equilibrium.

A density gradient can be achieved experimentally by generating a concentration gradient of sucrose or CsCl in a centrifuge tube, with the highest concentration being at the bottom and the lowest concentration at the top of the tube. The macromolecule under investigation will sediment in this gradient during centrifugation in a region where the density equals its own density. The density at this point is called the buoyant density. Macromolecules in a region of higher density will begin to float and migrate towards the area where the solvent and macromolecule densities are equal.

Historically, this method has delivered valuable information about nucleic acids, as it is highly sensitive to very small density differences. Contemporary applications of this method mostly use preparative ultracentrifugation, as this allows one to isolate preparative quantities of the separated species.

4.2 Mass spectrometry

Mass spectrometry is an analytical technique that provides information on the mass-to-charge ratio of charged particles. The technique has wide applications in a science laboratory as it can aid in:

- identification of unknown samples;
- structure elucidation of small molecules and proteins; and
- quantitative and qualitative analysis of samples.

Particular strengths of this technique are high sensitivity and accuracy, as well as the ability to interface with a range of separation (chromatographic) techniques. As only small amounts of a sample are required, mass spectrometry is an ideal method for all applications where the availability of the substance of interest is limited, for example in the area of environmental contaminant identification, forensic toxicology, proteomics and natural products research.

4.2.1 General schematics of a mass spectrometer

A mass spectrometer is composed of three main components: the ion source, an ion analyser and a detector. A general schematic of an instrument is depicted in Fig. 4.6.

The ion source generates ionised particles. Depending on the ionisation technique, the particle formed in the ion source will either be the charged intact parent structure, called the molecular ion, or structural fragments of the original molecule arising from the dissociation of chemical bonds due to the energy impact from the source. The mass analyser separates the ions in either space or time, based on their mass-to-charge ratio (m/z). The detector counts the ions and converts the electrical energy into a digital signal. There are a number of different ionisation methods and detectors used, with their applicability measured against the kind and type of molecule of interest for the analysis. A brief summary of the most common mass spectrometer components is given in the following sections.

4.2.2 Ionisation techniques

The development of new ionisation techniques constantly expands the range of molecules amenable to analysis by mass spectrometry. Table 4.2 lists the main ionisation techniques covered in this chapter and outlines their practical uses.

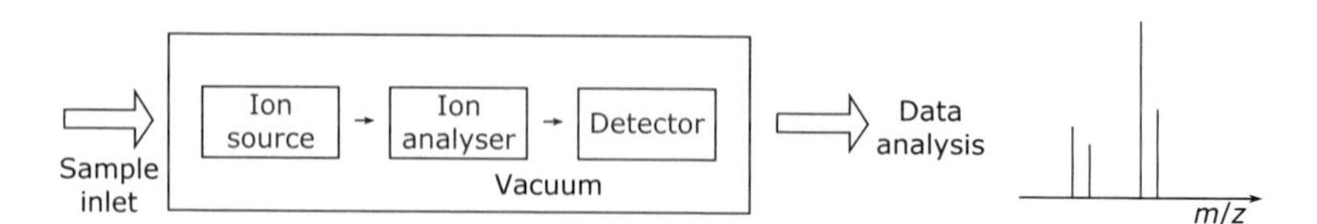

Fig. 4.6. General schematics of a mass spectrometer.

Table 4.2. Summary of popular ionisation techniques in mass spectrometry.

Ionisation technique[a]	Mass range of analytes	Applicability	Physico-chemical properties of analytes	Ions formed
EI	1500 Da	Small molecules	Non-polar, volatile	Odd electron ions, e.g. $[M^{\cdot}]^+$; extensive fragmentation
CI	1500 Da	Small molecules	Non-polar, volatile	Even electron ions, e.g. $[M+H]^+$; adducts, few fragments
FAB	5000 Da	Proteins, peptides, carbohydrates, organometallics	Polar, ionic, non-volatile	Even electron ions, e.g. $[M+H]^+$; matrix, adducts, few fragments
MALDI	300 000 Da	Proteins, peptides	Polar, ionic, non-volatile	Even electron ions, e.g. $[M+H]^+$, matrix
ESI	70 000 Da	Proteins, peptides, small molecules	Polar, ionic, non-volatile	Even electron ions, e.g. $[M+H]^+$; adducts, multiply charged
APCI	1500 Da	Small molecules	Non-polar, non-volatile	Even electron ions, e.g. $[M+H]^+$; few fragments

[a] See text for abbreviations.

Electron impact ionisation (EI)

Electron impact ionisation is the classical ionisation method in mass spectrometry. Under high vacuum (10^{-7} Torr = 10^{-5} Pa) and high temperatures (200 °C), the sample is thermally vaporised and the gas molecules are bombarded with a high-energy electron beam (70 eV) producing singly charged ions $[M^{\cdot}]^+$ that contain one unpaired electron. The amount of energy introduced into molecules by far exceeds their ionisation energy (typically around 15 eV) and often causes bonds to break, which, in turn, leads to subsequent rearrangements in the molecule. Ionisation methods that cause covalent bond breakage and formation of fragment ions are called 'hard'. In contrast, 'soft' ionisation methods provide ions without chemical bond cleavage. In hard ionisation techniques, multiple fragment ions may be produced and the initial molecular ion may not be detected. However, for standardised conditions (temperature of the ion source, energy of the electron beam), the number and the relative size of the fragment ions are reproducible and predictable for particular functional groups such as alcohols and esters; for example:

Alcohols $R\text{-}CH_2\text{-}OH \rightarrow [R\text{-}CH_2\text{-}H_2O]^+$

Esters $R_1\text{-}CH_2\text{-}COOR_2 \rightarrow [R_1\text{-}CH_3COOH]^+$

The EI spectra are thus characteristic for each substance and can be used in comparative searches. As such, they are amenable to library comparisons, automated mining of databases and interpretation, which makes them extremely useful in the analysis of unknown samples. The National Institute of Standards and Technology (NIST) database holds reference EI spectra of over 200 000 small molecules (http://chemdata.nist.gov/) and identification of known compounds often takes only seconds.

The EI technique is restricted to volatile, non-polar, low-molecular-mass substances (up to approx. 1500 Da); even air- and moisture-sensitive substances can be subjected to EI mass spectrometry. The analytes of interest therefore must be volatile and chemically stable to heat. This ionisation technique favours small, non-polar molecules; analysis of high-molecular-mass biomolecules is rather difficult. For higher throughput and automation, the technique can be coupled with gas chromatography.

Chemical ionisation (CI)

In the CI technique, the sample is introduced directly into ionised reagent gas, most commonly methane, isobutane or ammonia. Bimolecular collisions between sample molecules and gas lead to the ionisation of the sample. Due to the relatively low-energy transfer of this approach, this technique is significantly 'softer' compared with EI and thus results in lower molecular fragmentation and increased occurrence of molecular and adduct ions. As the delivery of the sample requires a moderate vacuum (1 Torr = 10^2 Pa) and high temperatures (200 °C), this technique is applicable to a similar range of molecules as EI, but with a higher probability of detecting the molecular ion of the analyte of interest.

Fast atom bombardment (FAB)

Fast atom bombardment is a desorption ionisation technique where the sample is mixed with a matrix and the mixture is desorbed from the surface upon bombardment by heavy particles. The sample is introduced under vacuum via a metal target coated with a liquid matrix containing the dissolved analyte. The matrix is composed of a non-volatile liquid solution such as *m*-nitrobenzyl alcohol, glycerol and thioglycerol (Fig. 4.7), which is then evaporated by a high-energy primary incident beam composed of Xe^+ or Cs^+ ions (5–20 keV) and thereby ionised. The sample molecules are desorbed from the surface, become charged by ion transfer and accelerate through to the analyser. Although FAB spectra show few fragments, the results are complicated by pseudo-molecular ions $[M+H]^+$ and adducts such as $[M+Na]^+$, $[M+Cs]^+$ and $[M+H+153]^+$ (nitrobenzyl alcohol matrix adduct). In addition, mono-charged non-covalently linked dimeric and

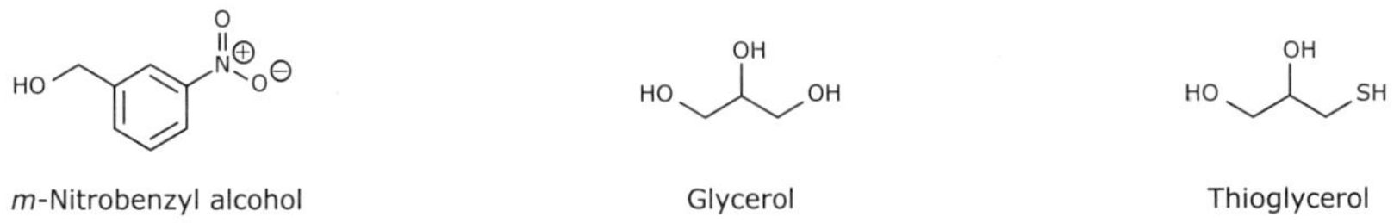

Fig. 4.7. Organic compounds used as the matrix in FAB experiments.

oligomeric artifactual ions are observed, which are barely distinguishable from the sample-specific covalently bound dimers and oligomers. Typical samples suited for FAB analysis are matrix-soluble, polar and non-volatile substances, which includes nucleotides, peptides, sugars and ionic and metallo-organic compounds. The practical molecular mass range of the technique is up to 5000 Da, which limits its applicability in the use of the analysis of proteins. Currently, it has largely been replaced by other more sensitive 'soft' ionisation methods such as MALDI and ESI (see below).

Matrix-assisted laser desorption/ionisation (MALDI)

In contrast to the other ionisation techniques covered so far, MALDI is a pulsed ionisation technique. The sample is co-crystallised with a matrix, most commonly consisting of UV-absorbing weak acids such as α-cyano-4-hydroxycinnamic acid, 3,5-dimethoxy-4-hydroxycinnamic acid and 2,5-dihydroxybenzoic acid (Fig. 4.8) and incorporated onto a metal plate. In contrast to FAB where the beam is composed of heavy atoms and ions, the MALDI primary incident beam is a nitrogen laser that emits light at 337 nm. The ionisation is achieved by a short laser pulse where the energy is absorbed by the matrix and then transferred to the analyte. Most of the high energy is absorbed by small molecules comprising the matrix, which results in the photo-ionisation of the matrix molecules. This, in turn, ionises the sample, yielding $[M+H]^+$ ions. MALDI is a 'soft' ionisation technique that allows the detection of molecular masses of the analyte with almost no fragmentation. The technique has a practical molecular mass range up to 300 000 Da and sensitivities in the order of femtomole amounts (10^{-15} moles). It is therefore ideally suited for the analysis of large biologically relevant molecules such as peptides, proteins, oligonucleotides, carbohydrates and lipids. Some of the difficulties associated with this technique are matrix interference for molecular masses of less than 700 Da, the possibility of photodegradation from the laser beam and the inability of the method to be combined with online liquid

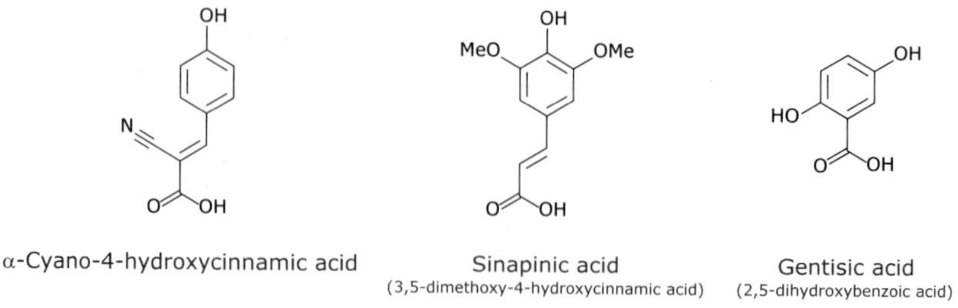

α-Cyano-4-hydroxycinnamic acid

Sinapinic acid
(3,5-dimethoxy-4-hydroxycinnamic acid)

Gentisic acid
(2,5-dihydroxybenzoic acid)

Fig. 4.8. Organic compounds used as the matrix in MALDI experiments.

chromatography. MALDI ionisation is most suited for large, polar, non-volatile macromolecules, especially in proteomics research.

Electrospray ionisation (ESI)

Electrospray ionisation is one of the most popular and commonly used interfaces in mass spectrometry today. The remarkable versatility and usefulness of this method is due to the ability to analyse molecules directly from the liquid phase at atmospheric pressure. The sample is delivered to the instrument in solution, most commonly methanol, acetonitrile, H_2O or combinations thereof, at very low flow rates. Weak organic acids, such as acetic acid or formic acid, are added to the solute in order to encourage ionisation of the sample. Gas-phase ions are generated by spraying the analyte solution from the tip of a capillary charged to a high voltage (2–5 kV). The charged droplets are reduced in size and transformed to ions by passing through a potential and pressure gradients en route to the analyser part of the instrument. The technique is 'soft', with very little fragmentation. The ESI spectra show adduct ions such as $[M+H]^+$ or $[M+Na]^+$, dimers $[2M+H]^+$ and multiply charged species. The appearance of these quasi-molecular ions enables determination of the molecular mass of large proteins, even when the molecular ion is outside the practical mass range of the instrument. Thus, ESI is suited for the analysis of peptides, proteins and small molecules.

Nano-electrospray ionisation (nanoESI) is a variant of ESI where the solvent delivery flow rates have been drastically reduced (nl min^{-1} in nanoESI, compared with μl min^{-1} for conventional ESI). The spray needle diameter is thus reduced and the tip of the needle positioned very close to the analyser. Therefore, this technique requires less sample for analysis and enhances the sensitivity into the range of low femtomoles (10^{-15} moles). The technique is particularly important in the analysis of biologically relevant macromolecules where the amount of sample is limited.

Atmospheric pressure chemical ionisation (APCI)

Similar to ESI, APCI generates ions directly from solution, whereby the sample is introduced together with nebulising gas into a heated vaporiser (400 °C). Ionisation is achieved by the aerosol particles passing by a discharge electrode where the solvent becomes ionised (corona discharge). The solvent ions then ionise analyte molecules, which is why this ionisation method is called 'chemical'. As the technique requires no vacuum ('atmospheric pressure'), the ionisation reaction between the solvent ions and the analyte is very efficient. Thus, APCI is an ionisation technique most suited for non-polar small molecules that would not usually ionise under other 'soft' methods such as ESI and MALDI.

Other ionisation techniques

Several other ionisation techniques are becoming increasingly popular and have been introduced in biomolecular science recently. Atmospheric pressure photo-ionisation (APPI) is a soft ionisation technique where the analyte is ionised upon exposure to UV light. Much like APCI, APPI is applicable to non-polar and non-volatile small molecules.

Direct analysis in real time (DART) is an ambient ionisation technique able to sample gas, liquid and solid samples. Here, ionisation is achieved by helium gas flowing through a tube divided into several chambers. The discharge chamber with electrodes holds a potential of 1–5 kV, causing ionisation of the analyte.

Surface-assisted laser desorption/ionisation (SALDI) is a technique similar to MALDI. However, the matrix used for SALDI consists of a liquid/graphene mixture, which reduces effects on the sample from matrix interference. Thus, SALDI has been proven a useful ionisation technique for the analysis of polar, non-volatile small molecules.

4.2.3 Analyser

The analyser separates ions based on their mass-to-charge ratio (m/z). This is achieved by applying electric and magnetic fields to the ions, accelerating them to a high velocity and separating them either in space (quadrupole analysers) or time (time-of-flight analysers).

The motion of a charged particle in vacuum can be described by two forces acting on it: the force accelerating the particle in an electric and magnetic field (Lorentz force), and the kinetic force derived from the particle's mass and velocity (Newton's second law of motion):

$$\vec{F}_{\text{Lorentz}} = q(\vec{E} + v\vec{B}) \quad (4.16)$$

and

$$\vec{F}_{\text{Newton}} = m\left(\frac{d\vec{v}}{dt}\right), \quad (4.17)$$

where $\vec{F}$ is the force acting on the ion, q is the charge, $\vec{E}$ is the electric field, v is the ion velocity, $\vec{B}$ is the applied magnetic field, m is its mass and dv/dt is the acceleration. Because both $\vec{F}_{\text{Lorentz}}$ and $\vec{F}_{\text{Newton}}$ describe the same net force on the ion, the two forces are equal, yielding the differential equation:

$$\left(\frac{m}{q}\right)\left(\frac{d\vec{v}}{dt}\right) = \vec{E} + v\vec{B}, \quad (4.18)$$

which relates a property of the ion, the mass-to-charge ratio m/q, with the experimentally set parameters $\vec{E}$ and $\vec{B}$. Instead of using the charge q (with units of $[q] = 1\,\text{C} = 1\,\text{As}$), it is common to refer to the unitless charge state z instead.

Because of the dependence on the charge state z, the doubly charged ion $[M+2H]^{2+}$ will appear with an apparent mass equalling $[M+2]/2$, the triply charged $[M+3H]^{3+}$ will appear at $[M+3]/3$, etc.

Quadrupole

The quadrupole analyser (also called quadripole analyser) is the most commonly used mass analysis technology today; it is typically coupled to EI, ESI and APCI ionisation sources. It consists of four charged electrodes connected to a radiofrequency and a direct current potential (Fig. 4.9). The radiofrequency and direct current potentials in the quadrupole function as high and low pass filters, respectively, and consequently only ions of select mass-to-charge ratios m/z are able to reach the detector. The range of detectable mass-to-charge ratios extends up to $m/z \leq 3000$ Da.

Quadrupole ion trap

The quadrupole ion trap is a variation of the quadrupole analyser, where the ions are trapped in a three-dimensional space between three electrodes: a ring electrode and two hemispherical end-cap electrodes. The ring electrode is positioned symmetrically between the two end-cap electrodes. Radiofrequency potentials applied to these electrodes regulate the movement of ions

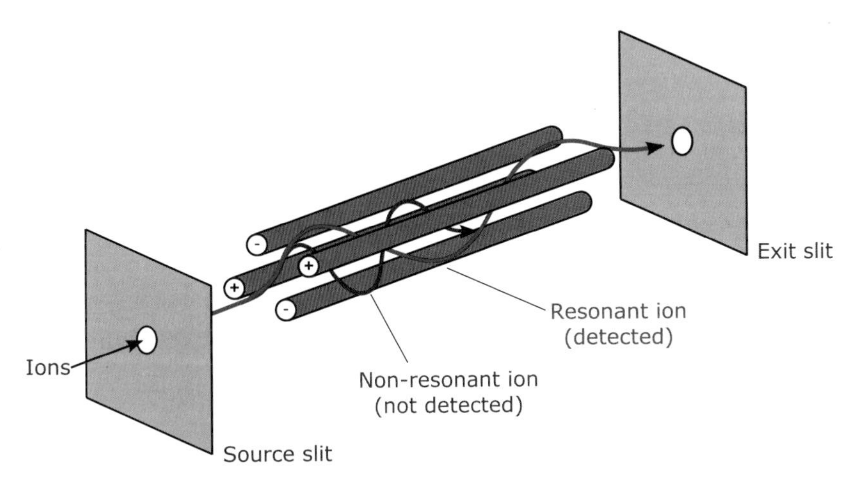

Fig. 4.9. Schematic of a quadrupole analyser. Each opposing cylinder pair is electrically connected. A radiofrequency with superimposed voltages is applied between the two pairs of cyclinders leading to selection of ions with a particular mass-to-charge ratio depending on whether they are resonant with the voltage ratios applied. Ions with unstable trajectories collide with the cylinders.

in and out of the trap. An advantage of the ion trap analyser is its small size and the ability to perform multiple-stage mass spectra (MS^n) experiments. For an MS/MS experiment, the ion of interest is selectively isolated in the trap, and fragmentation is achieved by resonant excitation of the target ion species. The analyser is often coupled to EI, ESI and MALDI ionisation techniques and is of particular importance in the analysis of protein sequences.

Time-of-flight (TOF)

The TOF analyser is a very popular feature of current mass spectrometers. The analyser is a flight tube of exactly known length and a triggered inlet grid. The observed parameter is the time needed for ions with a certain mass to pass through the entire flight tube. Ions are accelerated in an electric field whose energy can be calculated from $E = e \times U$. Due to the equivalence of electric and kinetic energies, all ions therefore have the same kinetic energy $E = {}^1/_2 mv^2$:

$$E = e \times U = \frac{m}{2} v^2. \tag{4.19}$$

Here, e is the charge of the electron and U is the voltage, while m is the mass of a particular ion and v its velocity.

The ions accelerate through a flight tube that uses a fixed voltage (2–25 kV). As the flight tube distance is fixed, selection will be for a set electric energy. All ions will have the same kinetic energy due to the above equation. However, their different masses will require different velocities, according to Newton's second law of motion. Ions with a low mass-to-charge ratio m/z therefore reach the detector first (they have high velocity), while ions with a high m/z will have lower velocity and take longer to travel through the flight tube. An advantage of this technique is the capacity to detect all ions in a sample and an unlimited mass range. The method is ideally suited for coupling with a MALDI ionisation source; MALDI-TOF instruments are at the forefront of proteomic research.

Tandem mass spectrometry analysis

Tandem mass spectrometry analysis (MS/MS) uses a combination of two detectors in order to achieve fragmentation data from select molecular ions. The initial stage of the analysis is the selection and isolation of a molecular ion of interest, followed by selective fragmentation. The most common systems in use are the triple quadrupole and the quadrupole/TOF (Q-TOF) instruments. Unlike the quadrupole ion trap analyser, where the ions are separated in time, here the ions are separated in space.

Triple quadrupole MS/MS systems operate in the following manner: quadrupole 1 (MS1) selects and isolates a single ion from a mixture and the second quadrupole acts as a collision cell where fragments are generated, while the third quadrupole (MS2) separates the fragments according to their mass-to-charge ratio m/z. In a Q-TOF instrument, the third quadrupole is replaced by a TOF analyser.

Fourier transform ion cyclotron resonance (FT-ICR, FT-MS)

The method of FT-ICR or FT-MS is the most sensitive analyser method in use today, and can achieve an accuracy of less than 1 ppm and a resolution of 10^9 Da. In an FT-MS analyser, the ions are exposed to a static and uniform magnetic field and will move in a circular motion perpendicular to the direction of the applied field. The angular frequency of an ion's orbit is called the cyclotron frequency (ω) and is determined by the strength of the magnetic field (B) and the mass-to-charge ratio m/z of the ion:

$$\omega = \frac{B \times e}{m/z}. \tag{4.20}$$

When a radiofrequency pulse sweep is applied to the transmitter electrodes, it will produce a cyclotron frequency signal for all of the ions present. The resulting signal is similar to a free induction decay of a nuclear magnetic resonance (NMR) experiment (see Section 3.2.1). Using Fourier transformation, the signal can be converted from the time domain to the frequency domain and thus provide a complete mass spectrum for the sample. A detector is therefore not needed in FT-MS. However, the high resolution provided by an FT-MS instrument comes at a higher cost compared with other mass spectrometric techniques. As it requires a high-field magnet for operation, maintenance costs and bench space for this type of instrument are relatively high, which explains why it is not as prominent as other mass spectrometers.

Other analyser types

Like many other techniques, the development of mass spectrometry has been intimately linked to technological advances. Analysers of historical importance include:

- Static magnetic analyser: This analyser detects all ion masses simultaneously because the ions generated in the ion source simply fly through a space with a surrounding permanent magnet. The magnetic field scatters the ion beam such that light ions are diverted much more than heavy ions, and thus ions with different masses each hit the detection plate at a separate position. In earlier times, the detection plate was a photographic film that had to be analysed by an optical densitometer. Nowadays, this type of analyser is coupled with a diode array. The simultaneous detection of the whole mass spectrum enables extremely short running times.
- Separation tube: Mass discrimination with a separation tube requires the ion to be detected to fly through an angled tube while ions that are lighter or heavier collide with the tube and therefore are prevented from flying through. Typical tube angles range from 30 to 60°.
- Reflectron: This is a variation of a TOF analyser. An ion mirror is used to reverse the direction of the travelling ions by applying a retarding electric field. This addition to the TOF analyser can be used to improve mass resolution.

4.2.4 Detection techniques

Traditional analogue (Faraday cup)

A Faraday cup detector is composed of an individual cylindrical dynode (electrode), which, upon contact with ions, will emit secondary electrons and induce a current. The current can be detected and measured to give a mass spectrum. The detector is not very sensitive and as such is not the preferred detector technique currently used.

Electron multiplier (EM)

An EM is the most common detector technology used today. Instead of a single dynode, an EM has a series of dynodes held at increasing potentials. The ions striking the first dynode produce secondary electrons, which cascade and multiply with each successive dynode contact, resulting in an overall amplification of signal and current gain. For each electron striking the initial amplification dynode, typically the overall current gain is 1 million to one.

4.2.5 Applications of mass spectrometry

Small molecules

The ability to couple mass spectrometers to liquid chromatography separation techniques such as HPLC (high-performance liquid chromatography) and UPLC (ultrahigh-performance liquid chromatography) has made mass spectrometry the most impotant analytical tool in the analysis of small molecules today. The uses and applications are comprehensive and include:

- identification and structure elucidation of natural products;
- purity and quantitative analysis of mixtures;
- study of drug metabolism and pharmacokinetics;
- trace contaminant determination in environmental science; and
- forensic toxicology.

Macromolecules

Mass spectrometry is an essential tool for the study of biologically relevant macromolecules largely due to the introduction of 'soft' ionisation techniques such as ESI and MALDI, providing the ability to ionise high-molecular-mass compounds such as proteins, oligonucleotides and oligosaccharides. Frequently, protein samples are subjected to mass spectrometric analysis after gel electrophoresis, which enables investigation of individual protein species without having purified individual samples. Multiple charging observed in the ESI spectra of large molecules containing basic groups enables the determination of the molecular mass of proteins, even when the mass of the molecular ion is outside the nominal range of the instrument. For two different protonation states of a multiply charged protein with adjacent peaks *X1* and *X2* differing by a single charge, the molecular mass (*M*) can be determined as follows:

$$(m/z)_{X1} = \frac{M + n \times M(\mathrm{H}^+)}{n \times M(\mathrm{H}^+)} \quad (4.21)$$

is the mass-to-charge ratio of peak *X1*, and

$$(m/z)_{X2} = \frac{M + (n+1) \times M(\mathrm{H}^+)}{(n+1) \times M(\mathrm{H}^+)} \quad (4.22)$$

is the mass-to-charge ratio of peak *X2*.

Resolving both equations for the charge n on the molecule observed at peak *X1* yields:

$$n = \frac{(m/z)_{X2} - 1 \times M(\mathrm{H}^+)}{(m/z)_{X1} - (m/z)_{X2}}. \quad (4.23)$$

The charge of the molecule observed at peak *X1* can now be inserted into the first equation, resolved for the molecular mass:

$$M = n \times [(m/z)_{X1} - 1 \times M(\mathrm{H}^+)]. \quad (4.24)$$

For example, if peaks are observed at *X1* with a m/z ratio of 1431.6 Da, and at *X2* with a m/z ratio of 1301.4 Da, then the charge n at peak *X1* is calculated as:

$$n = \frac{1301.4\ \mathrm{Da} - 1 \times 1.008\ \mathrm{Da}}{1431.6\ \mathrm{Da} - 1301.4\ \mathrm{Da}} = 10. \quad (4.25)$$

This yields a molecular mass for the molecule of:

$$M = 10 \times (1431.6\ \mathrm{Da} - 1 \times 1.008\ \mathrm{Da}) = 14305.9\ \mathrm{Da}. \quad (4.26)$$

Another important application of mass spectroscopy in the study of peptides and proteins is amino acid sequencing by tandem mass spectrometry. Following enzymatic digestion, the peptide mixture is chromatographically separated (usually on a reversed-phase HPLC column) and each constituent peptide analysed by tandem mass spectrometry. The fragmentation of peptides occurs along the peptide bond to give b and y ions originating from the amino- and carboxyl-termini, respectively (Fig. 4.10). The sequential losses of amino acids in the mass spectra will give a specific sequence of b and y ions, from which the amino acid sequence of particular peptides can be worked out.

In a related approach, called MS fingerprinting, the protein sample is subjected to enzymatic digestion and the resulting mixture of peptides is

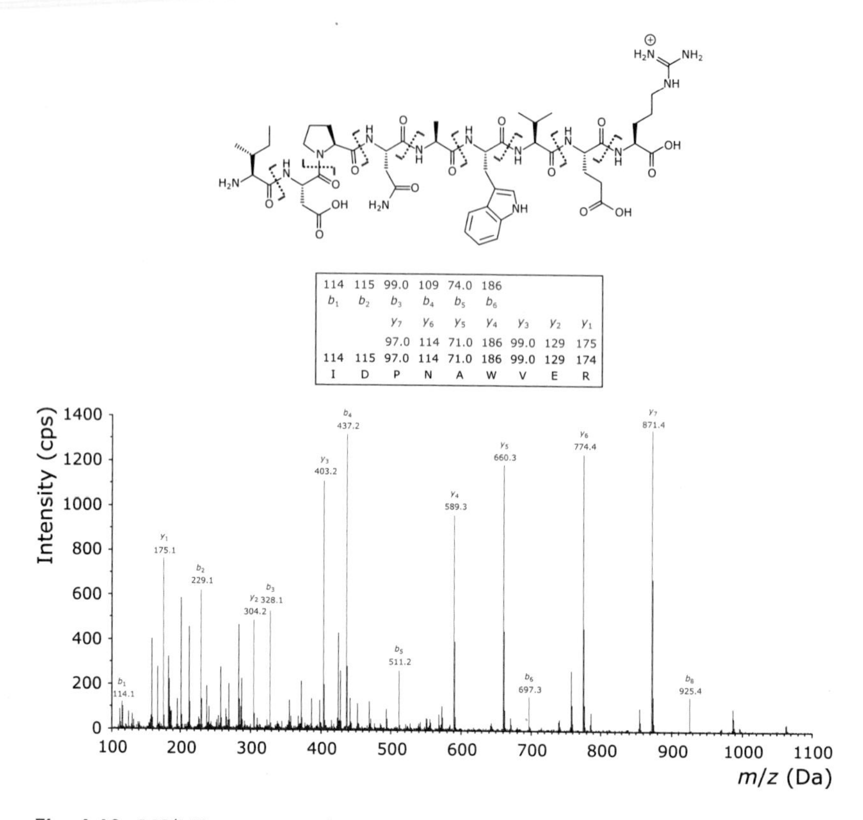

Fig. 4.10. MS/MS spectrum of a peptide fragment derived from *Escherichia coli* β-galactosidase. (Data courtesy of Dr Michelle Colgrave, CSIRO.)

analysed by mass spectrometry, typically MALDI. The data obtained contain masses for various peptide fragments that are characteristic for a given protein. The resulting spectrum can be used for computational spectral library matching. Individual proteins can thus be identified from databases generated either from experimental data or by *in silico* digestion of amino acid sequences of proteins.

4.3 Calorimetry

Physico-chemical changes in matter are accompanied in almost all cases by energetic changes. An important parameter in this context is the internal energy U of a system, which is defined as the sum of all energies intrinsic to a thermodynamic system. With solids, and most importantly in the area of the life sciences, liquids or solutions, changes of the internal energy U are best analysed under constant pressure, i.e. allowing for volume changes during the processes to be investigated. For such processes, the enthalpy H has been defined as a matter of convenience to describe the energy function of solid and liquid systems:

$$H = U + pV. \tag{4.27}$$

Upon differentiation, this yields

$$dH = dU + pdV + Vdp = dQ + dW + pdV + Vdp, \tag{4.28}$$

where dQ describes heat energy, and dW, pdV and Vdp describe the work energy of the system. The term Vdp will be zero, because we established that calorimetry of solids and liquids is to be carried out under constant pressure, i.e. $dp = 0$. If one further assumes that all other types of work can be excluded, then one can suppose that $dW = 0$. Therefore, the enthalpy change ΔH can be expressed as:

$$dH = dQ + pdV. \tag{4.29}$$

For systems with negligibly small volume changes during the experiment, the assumption $dV=0$ applies, and the enthalpy can be determined through the heat energy:

$$\Delta H = \Delta Q. \tag{4.30}$$

Based on the enthalpy, a material-specific constant called the heat capacity C_p can be defined. This parameter describes the changes observed in the enthalpy upon changing the temperature of the system (under constant pressure):

$$C_p = \frac{\Delta H}{\Delta T}. \tag{4.31}$$

Calorimetric measurements record the heat changes of thermodynamic systems either in the process of a changing system composition or during environmental changes of a system of given composition. As discussed above, in many cases this enables direct determination of the enthalpy. Importantly, most other biophysical approaches measure changes in the free energy (Gibbs' energy):

$$\Delta G = \Delta H - T\Delta S. \tag{4.32}$$

Certain effects that provoke a change in enthalpy might not be visible from changes in the free energy, because of counteracting entropy changes. With proteins, this phenomenon of enthalpy–entropy compensation is well known (Dunitz, 1995). The ability of calorimetry to measure enthalpies (not free energies) directly makes it unique, as it is the only experimental methodology that allows direct determination of enthalpies.

Due to the ubiquitous occurrence of enthalpy (heat) changes in chemical and biochemical reactions or processes, calorimetry has established itself as a versatile tool because measurements can be made:

- in solution;
- without the requirement for spectroscopically clear samples;
- without the requirement for immobilisation (e.g. in surface plasmon resonance); and
- without the requirement for conjugation with reporter groups (e.g. fluorophores).

There are two commonly used experimental approaches for calorimetric experiments. One method measures the heat released by a system upon addition of portions of reaction components (titration calorimetry). The other monitors heat coming out of a given system while gradually changing the temperature of the system (scanning calorimetry).

4.3.1 Isothermal titration calorimetry (ITC)

Molecular recognition is a fundamental process in enzymology and the underlying thermodynamics are of enormous interest and are necessary to characterise and understand ligand binding fully. Isothermal titration calorimetry has become a routine method for assessing thermodynamic data of association (binding) processes. It detects heat changes during physico-chemical processes and thus offers a broad range of applications in reactions involving enthalpy changes, including chemical and biochemical binding studies as well as enzyme kinetics. It is a direct, extremely sensitive method, and does not require chemical modification or immobilisation of the sample of interest. Particular applications in the life sciences include:

- protein–ligand binding;
- protein–protein binding; and
- reaction kinetics.

However, more generally, any compound–compound interaction can be assessed, i.e. small-molecule systems are equally amenable to this technique.

Instrumentation

Isothermal titration calorimeters (Fig. 4.11) consist of two identical cells, which are made of highly efficient thermal conducting material such as Hasteloy or gold. The cells are surrounded by an adiabatic jacket to ensure an accurate accounting of all heat. This is achieved by thermocouple

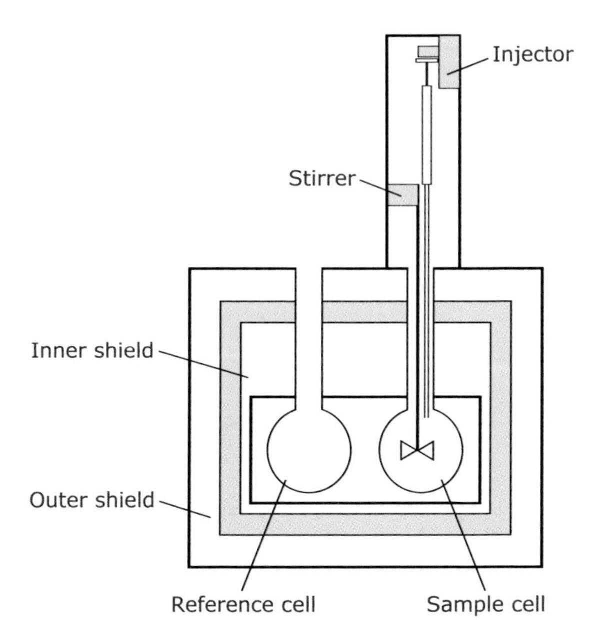

Fig. 4.11. Schematics of an isothermal titration calorimeter.

circuits detecting any temperature differences between the sample and the reference cells, and between the cells and the jacket. Heaters situated on both cells as well as the jacket are activated as required to maintain the same temperature with all components (hence the term isothermal calorimeter). Note that no adiabatic device is truly adiabatic; in ITC, some heat will be transferred between the sample and the sample holder and thus be unaccounted for during the measurement. Some instruments thus consider a correction factor to account for this heat loss. The correction factor is calculated as the ratio of the thermal mass of the sample and sample holder to the thermal mass of the sample. A reaction component is delivered into the sample cell by means of an automated syringe mechanism that injects precise volumes at specific time intervals.

Measurement principle

The calorimeter measures the rate of heat flow, which results from heat changes due to a reaction in the sample cell. As described above, the instrument therefore monitors the differential heat effects between the sample and the reference cell by a feedback mechanism. A constant power is applied to the reference cell, and the temperature difference between

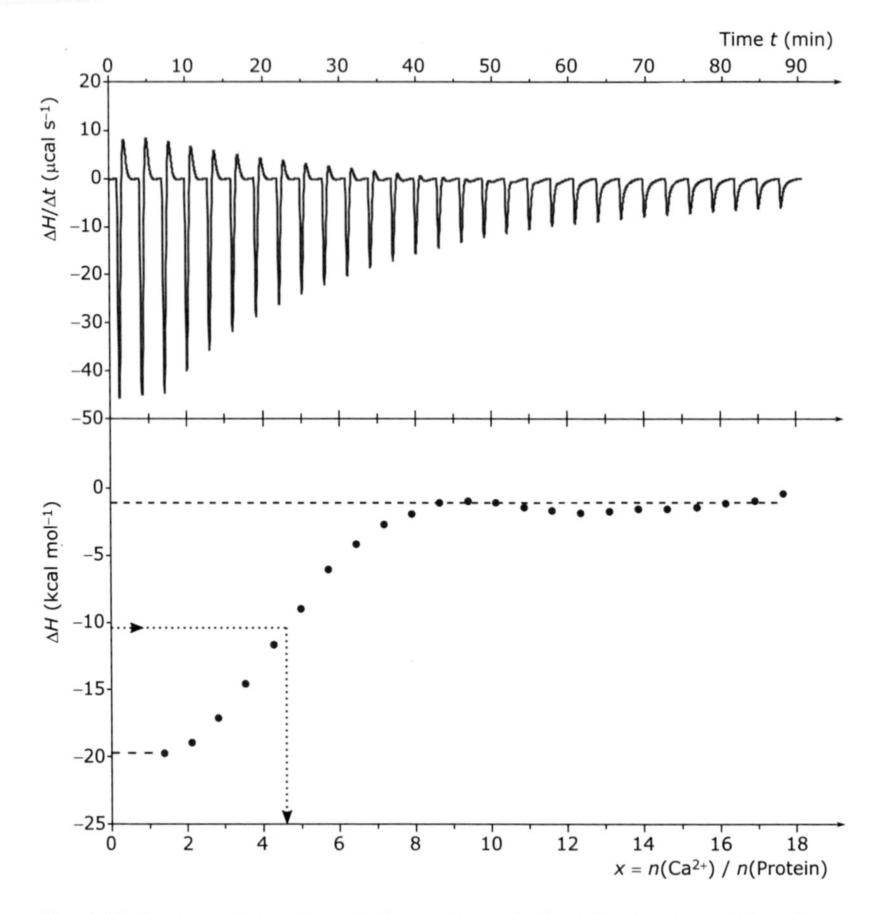

Fig. 4.12. Isothermal titration calorimetry data obtained for titrating calcium ions into a solution containing the protein annexin (Gh1). (Data courtesy of Dr Nien-Jen Hu). Top: Raw data. The area underneath each injection peak is equal to the total heat released for that injection. Bottom: Plot of the integrated heat against the molar ratio of ligand added to protein. The dotted and dashed lines illustrate determination of stoichiometry at the point of inflection of the binding isotherm (in this example approximately five bound calcium ions (Ca^{2+}) per annexin monomer).

both cells is monitored. A temperature difference between both cells causes application of a variable power to the sample cell (feedback power) in order to maintain the same temperature in both cells. This feedback power is what is recorded by the ITC instrument.

The feedback power in the absence of any reaction in the sample cell represents the baseline level. Exothermic reactions in the sample cell will increase the heat in the sample cell and lead to a decrease in feedback power. Contrarily, endothermic reactions will absorb heat from the sample cell and thus cause an increase in feedback power.

As shown in Fig. 4.12, the raw data, which is the recorded feedback power (heat per time) in response to an injection, needs to be integrated to obtain the actual heat released or absorbed. In other words, the area enclosed by the baseline deviations (spikes) of ITC data is a measure of the heat (and therefore the enthalpy) evolved or consumed per reaction.

Experimentation

For illustration, we consider a protein–ligand binding experiment. While ITC can be used to determine an enthalpy change (ΔH) by a single injection, it is most frequently performed as a titration series covering a range of molar ratios between ligand and protein, thus allowing analysis of association constants and binding stoichiometry. Therefore, an ITC experiment consists of injections of ligand portions into a protein solution or possibly the other way round. The observed heat change comprises effects from several sources:

- heat of binding (to be measured);
- heat of dilution of protein;
- heat of dilution of ligand; and
- heat of mixing.

Consequently, there may be three control experiments to be performed in order to correct for the heat effects due to dilution and mixing. However, by restricting dilution of the initial protein solution to about 20%, the heat changes due to protein dilution are generally taken as negligible. Similarly, the heat of mixing is usually small and can be neglected, if all buffers are matched, i.e. all buffers used are from the same batch. The dilution of ligand is almost always of major significance, as the starting concentration in the calorimeter cell is zero and aliquots of ligand are added from a high concentration in the injection syringe until the final concentration in the cell is several times that of the final protein concentration. Thus, the heat change due to ligand dilution cannot be neglected, and a control experiment of ligand titration into buffer should be performed. The measured heat of dilution then has to be subtracted from the heat effect measured in the main experiment.

Alternatively, subtraction of a linear regression through the last few data points of titration can often be used as an approximation for the heat effects due to dilution and mixing, without the need for a further control experiment (Holdgate, 2001).

Characterisation of binding

Binding experiments are performed by titrating protein with ligand portions added in individual injections. Ligand subsequently added to the cell will bind to the available protein-binding sites and the heat change will depend on the molar number of complexes formed according to the reaction:

$$\text{Protein} + n\ \text{Ligand} \underset{K_d}{\overset{K_a}{\rightleftharpoons}} [\text{Protein} : \text{Ligand}_n]$$

Determination of reliable parameters depends on correct experimental design and analysis. The parameter C is a unitless parameter that describes the product of association constant K_a, protein concentration c and stoichiometry n:

$$C = K_a \times c(\text{Protein}) \times n \tag{4.33}$$

In order to have free ligand and complex at significant concentrations within the experiment, and thus allow for determination of reliable values for K_a and n from the binding isotherm, C values between 5 and 100 are recommended. This can in many cases be achieved by choosing protein concentrations of around the value of the dissociation constant of the complex, K_d ($= {K_a}^{-1}$). While low C values yield almost horizontal isotherms, very large C values result in rectangular-shaped isotherms. In the former case, neither ΔH nor K_a can be determined, while in the latter case, information on ΔH and stoichiometry but not on K_a is available.

It is desirable to arrange experimental conditions so that all three parameters can be determined in a single titration experiment. Data points in the initial part of the titration series provide information on the magnitude of ΔH. Data points within the slope of the isotherm contain information about affinity and stoichiometry. The end region of the titration series holds information about mixing and dilution effects.

Assuming that the concentrations of protein and ligand are accurately known, the stoichiometry of complex formation is easily determined from the molar ratio of ligand and protein at the point of inflection of the isotherm (Fig. 4.12).

The limitations of binding characterisation with ITC are given by the (im)practicability of optimal concentrations. Very weak binding (low affinity) requires high protein concentrations, which may not be available or practical due to solubility or aggregation problems. Very strong binding (high affinity) would require extremely low protein concentrations, which would produce heat changes not measurable with the ITC instrument. Common instruments are thus limited to a lowest K_a of $10^9\,M^{-1}$ ($K_d = 1\,nM$).

Analysis of ITC binding data

The heat change upon binding in each injection in a calorimetric titration series is proportional to the change in concentration of ligand bound:

$$\Delta Q = V \times \Delta H \times \Delta c(\text{Ligand}_{\text{bound}}), \tag{4.34}$$

where V is the reaction volume and ΔH the binding enthalpy. The cumulative heat Q can be expressed in terms of c_0(ligand):

$$Q = \frac{[1 + nK_a c_0(\text{Protein}) + K_a c_0(\text{Ligand})] - \sqrt{[1 + nK_a c_0(\text{Protein}) + K_a c_0(\text{Ligand})]^2 - 4nK_a^2 c_0(\text{Protein}) c_0(\text{Ligand})}}{\frac{2K_a}{V\Delta H}} \tag{4.35}$$

Note that this equation is based on total concentrations c_0 of ligand and protein in the calorimetric cell, and thus enables determination of the three unknown parameters (K_a, n and ΔH) by means of curve fitting of the cumulative heat data from an ITC experiment.

Alternatively, the incremental heat signal ΔQ can be plotted against the molar ratio of ligand and protein and yields the familiar sigmoid titration curve (cf. Fig. 4.12; Wiseman *et al.*, 1989). Evidently, both methods should yield the same values for parameters analysed; any differences indicate the presence of systematic errors (Bundle & Siguskjold, 1994).

It is worth emphasising that ΔH is the apparent enthalpy of the binding reaction, as other heat changes might arise from linked equilibria, like for example large conformational changes of the protein upon binding. Similarly, the heat of ionisation of buffer must be taken into account if the binding reaction is associated with changes in protonation states.

As ΔH and K_a can be determined from ITC experiments, the values of the Gibbs' (free) energy, ΔG, and the entropy, ΔS, for the investigated reaction can be obtained from:

$$\Delta G = \Delta H - T\Delta S = -RT\ln K_a. \tag{4.36}$$

The above treatment is a simplification, as it is assumed that K_a as well as ΔH are independent of the temperature, which is rarely the case. A more accurate description of binding thermodynamics requires consideration of the heat capacity ΔC_p (specific heat), which itself normally is temperature dependent. Further discussion of these advanced analyses can be found in the literature (Cooper, 1999).

4.3.2 Differential scanning calorimetry (DSC)

Differential scanning calorimetry monitors the heat coming from or consumed by a sample while varying the temperature of the system at a constant rate. The main applications of this technique have been in materials research, but it is frequently employed in the life sciences for studying:

- protein folding and stability;
- protein–membrane interactions; and
- physical drug properties in the pharmaceutical industry.

Instrumentation

A DSC instrument comprises a sample pan and an empty reference pan that are placed on a thermoelectric disc made of constantan and embedded in a furnace, which is typically made from silver (Fig. 4.13). The furnace is heated at a constant temperature rate, and both pans receive the same heat input through the thermoelectric disc. Due to the heat capacity of the sample C_p, temperature differences between the sample and the reference pan will arise during the heating process. These are measured by sensitive devices called thermocouples;

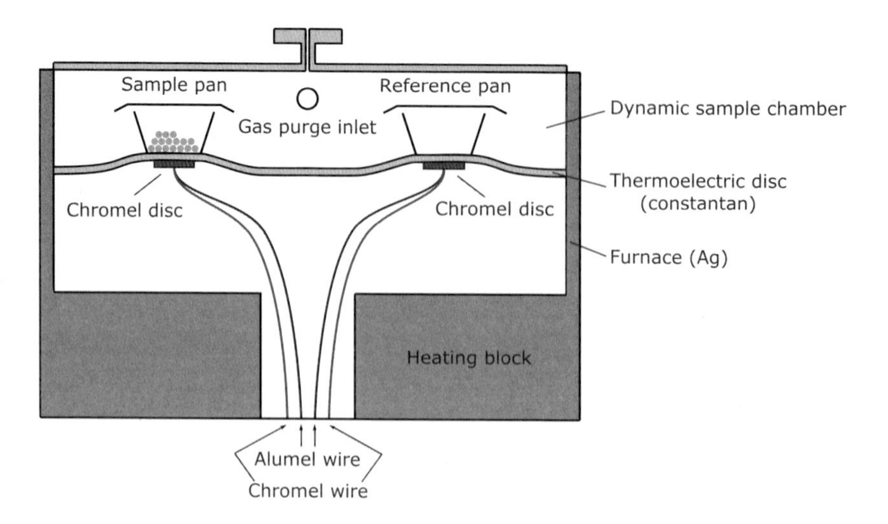

Fig. 4.13. Schematic layout of a differential scanning calorimeter.

thermocouples are electrodes made of two different alloys such as Ni-Cr ('chromel') and Ni-Al ('alumel'). The temperature difference ΔT between the sample and the reference pan is converted to a parameter called heat flow q, which is defined by analogy to Ohm's law, which relates electric current, voltage and resistance:

$$q = \frac{\Delta T}{R} \tag{4.37}$$

Here, R is the thermal resistance. The temperature difference ΔT is measured as the voltage difference at the junctions between the thermoelectric disc and the thermocouples on the sample and reference compartments. The voltage difference is then adjusted for the thermocouple response, and the resulting read-out is proportional to the heat flow.

The data obtained from a DSC experiment comprise heat flow versus temperature. As the heat flow originates from the sample, it can be assumed that $q = \Delta Q = \Delta H$. From the heat flow equation above, we thus obtain:

$$\frac{q}{\Delta T} = \frac{1}{R} = \frac{\Delta H}{\Delta T} = C_{\mathrm{p}}. \tag{4.38}$$

Experimentation

From the above, it is obvious that results from a DSC experiment can be plotted as heat capacity versus temperature (called a thermogram; see Fig. 4.14). From the curve obtained, exothermic and endothermic reactions in the sample can be distinguished, as they will appear as peaks of opposite directions (whether exothermic peaks are positive or negative depends on the particular instrument).

From such curves, one can determine the enthalpy of the process under investigation by integrating the peak of a particular transition (area under the curve, multiplied by a calorimetric constant that is to be determined for a particular instrument). The characteristic change of heat capacity can be determined from the difference between the heat capacities before and after the transition occurred.

Applications

The technique of DSC is well suited to studying the folding and stability of proteins and allows the determination of thermodynamic parameters associated with protein stability. Generally, the temperature of the

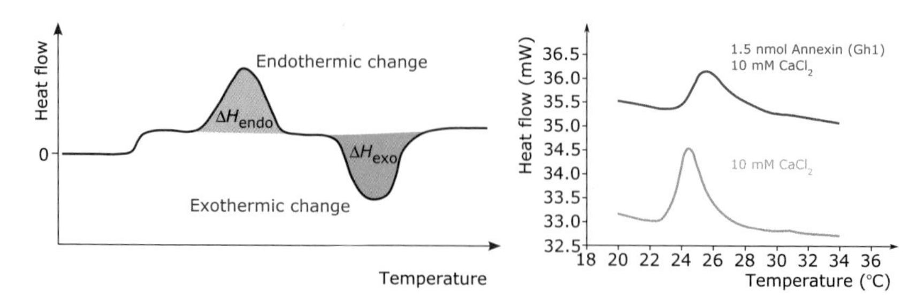

Fig. 4.14. Left: Schematic thermogram from DSC. The signal represents the change in heat capacity in the sample. Endothermic and exothermic changes give rise to opposing extrema in the thermogram. Right: DSC thermograms show the change of the transition of multi-lamellar vesicles from the gel to the liquid–crystalline state. In the presence of the membrane-binding protein annexin (Gh1) and calcium ions, the transition occurs at a higher temperature indicating a stabilising effect of the protein on the membrane. The protein binds to the membrane in a calcium-dependent manner. (Data courtesy of Dr Nien-Jen Hu.)

maximum of the C_p curve during a transition is interpreted as the 'melting temperature' T_m representing the stability of the sample (Cooper & McAuley, 1993).

In protein unfolding experiments, it is assumed that 50% of the protein is unfolded at the melting temperature T_m. Higher values of T_m indicate more stable species. The change in heat capacity during the unfolding process, ΔC_p, is believed to be due to changes in the hydration of side chains, which become increasingly exposed to solvent during the unfolding process. The enthalpy that is determined for a particular process by integration of the thermogram, frequently denoted as ΔH_{cal}, can be compared with the van't Hoff enthalpy, ΔH_{vH}, which is determined independently from the shape of the thermogram peak. Determination of the van't Hoff enthalpy assumes a thermodynamic model that is fitted to the experimental data. If ΔH_{cal} and ΔH_{vH} are in agreement, the underlying transition can be concluded to be a two-state transition.

For a particular system under study, the different transitions in a DSC recording need to be assigned their underlying molecular process, for which the relevant thermodynamic parameters can then be determined. In the example shown in Fig. 4.14, the effect of a protein on phospholipid membranes has been investigated.

In addition to basic research applications, DSC is frequently used as a routine method for characterisation of physical properties, especially in the development of polymers and materials, as well as the pharmaceutical industry. In the design of drug formulations, for example, the stability of drugs needs to be determined to assess their storage and shelf-life properties. Also, drugs are preferably delivered as amorphous substances, so during the production process compounds need to be handled at temperatures that avoid crystallisation.

FURTHER READING

Centrifugation

Mächtle, W. & Börger, L. (2006). *Analytical Ultracentrifugation of Polymers and Nanoparticles*. Berlin, New York: Springer Verlag.

Scott, D. J., Harding, S. E. & Rowe, A. J. (2006). *Analytical Ultracentrifugation: Techniques and Methods*. Cambridge, UK: RSC Publishing.

Websites on centrifugation

UltraScan sofware archives: http://structure.usc.edu/ultrascan/

SEDPHAT: global analysis of sedimentation velocity, sedimentation equilibrium and/or dynamic light-scattering experiments: http://www.analyticalultracentrifugation.com/sedphat/

Mass spectrometry

Cody, R. B., Laramée, J. A. & Durst, H. D. (2005). Versatile new ion source for the analysis of materials in open air under ambient conditions. *Analytical Chemistry* **77**, 2297–302.

de Hoffmann, E. (1996). Tandem mass spectrometry: a primer. *Journal of Mass Spectrometry* **31**, 129–37.

Gaskell, S. J. (1997). Electrospray: principles and practice. *Journal of Mass Spectrometry* **32**, 677–88.

Hunt, D. F., Yates, J. R. III, Shabanowitz, J., Winston, S. & Hauer, C. R. (1986). Protein sequencing by tandem mass spectrometry. *Proceedings of the National Academy of Sciences, USA* **83**, 6233–7.

Kind, T. & Fiehn, O. (2010). Advances in structure elucidation of small molecules using mass spectrometry. *Bioanalytical Reviews* **2**, 23–60.

March, R. E. (1997). An introduction to quadripole ion trap mass spectrometry. *Journal of Mass Spectrometry* **32**, 351–69.

Marshall, A. G., Hendrickson, C. L. & Jackson, G. S. (1998). Fourier transform ion cyclotron resonance mass spectrometry: a primer. *Mass Spectrometry Reviews* **17**, 1–35.

Pretsch, E., Bühlmann, P. & Badertscher, M. (2009). *Structure Determination of Organic Compounds*, 4th edn. Berlin, Heidelberg: Springer Verlag. (A compendium with tables of spectral data.)

Websites on mass spectrometry

Organic Chemistry at CU Boulder: Mass Spectroscopy: http://orgchem.colorado.edu/Spectroscopy/Spectroscopy.html
Scripps Center for Metabolomics and Mass Spectrometry: http://masspec.scripps.edu/index.php

Calorimetry

Jelesarov, I. & Bosshard, H. R. (1999). Isothermal titration calorimetry and differential scanning calorimetry as complementary tools to investigate the energetics of biomolecular recognition. *Journal of Molecular Recognition* 12, 3–18.
Velazquez-Campoy, A. & Freire, E. (2006). Isothermal titration calorimetry to determine association constants for high-affinity ligands. *Nature Protocols* 1, 186–91.
Wilcox, D. E. (2008). Isothermal titration calorimetry of metal ions binding to proteins: an overview of recent studies. *Inorganica Chimica Acta* 361, 857–67.

Websites on calorimetry

Calorimetry Technology Platforms: http://www.microcal.com/technology/
Isothermal Titration Calorimetry: Experimental Design, Data Analysis, and Probing Macromolecule/Ligand Binding and Kinetic Interactions (FEBS Practical Course 2009): http://www.enzim.hu/febs2009/pdfs/Lewis_ITC.pdf
Differential Scanning Calorimetry: http://pslc.ws/macrog/dsc.htm

Surface-sensitive methods

5

5.1 Surface plasmon resonance

Surface plasmon resonance (SPR) is a surface-sensitive method for monitoring the smallest changes of the refractive index or the thickness of thin films (Salzer & Steiner, 2004). In bioanalytical applications, where interactions between biomolecules are to be detected without the use of labels, SPR has been the method of choice for many years. Therefore, the most important applications are:

- monitoring of competitive binding;
- receptor studies;
- kinetic characterisation of chemical processes; and
- analysis of structural domains in thin layers and films.

There are many advantages of SPR compared with other biochemical techniques. Unlike other techniques, for SPR there is no need to label individual components, and measurements can be made using opaque/turbid samples. As the field wave, but not the light beam, penetrates the sample, there is no requirement for spectroscopically transparent samples. The detection limit is at the order of 10^{-12} M (pM) and the binding partners can be captured from a mixture (no purification required). Immobilised target molecules can be reused after washing with a base between interactions.

The main disadvantages of SPR are non-specific detection and sensitivity to temperature changes, but these can be compensated by modern techniques in image processing and constant sample temperatures.

Recently, SPR has also been used as an imaging method to characterise thin layers and films in an area-resolved manner. Thus, SPR imaging provides highly sensitive detection of changes in the refractive index or the thickness of layers.

Historically, SPR imaging was used only to investigate micro-structured layers in certain non-biological research areas. With the rapid progress in DNA and protein silicon chips in the 1990s, a label-free detection method was in high demand that could be satisfied by SPR imaging. Components such as laser diodes, fast and sensitive charge-coupled device (CCD) cameras, and filter and micro-optical elements became widely available and less expensive with the innovations happening in data transfer and telecommunication. Whereas basic SPR instruments use a linear photodiode array detector, two-dimensional array detectors enable the employment of SPR as an imaging technique.

5.1.1 Background

Plasmons are electromagnetic oscillations at the surface of metal nanoparticles in direct contact with a dielectric medium (such as glass) and arise when the particles are hit by incoming photons. Excitation of a plasmon wave therefore requires an optical prism with a metal film of about 50 nm thickness. Total internal reflection occurs when a light beam travelling through a medium of higher refractive index (e.g. a glass prism with a gold-coated surface) meets at an interface with a medium of lower refractive index (e.g. aqueous sample) at an angle above the critical angle.

Total internal reflection of an incident light beam at the prism–metal interface elicits a propagating plasmon wave by leaking an electrical field intensity, called an evanescent field wave, into the medium of lower refractive index. As the interface between the prism and the medium is coated with a thin layer of gold, incident photons excite a vibrational state of the electrons of the conducting band of the metal. In thin metal films, this propagates as a longitudinal vibration. The electrons vibrate with a resonance frequency that is dependent on the metal and prism properties, as well as on the wavelength and the angle of the incident beam. Excitation of the plasmon wave leads to decreased intensity of the reflected light. Thus, SPR produces a dip in the reflected light intensity at a specific angle of reflection (Fig. 5.1).

The propagating surface plasmon wave enhances the amplitude of the evanescent field wave, which extends into the sample region. When binding occurs, the refractive index on the sample side of the interface increases. This alters the angle of incidence required to produce the SPR effect and hence also alters the angle of reflected light. The change in angle brings about a change in intensity recorded at the detector, which can be plotted against time to give a sensorgram reading (Fig. 5.1).

Because in SPR instruments the angle α and the wavelength of the incident beam are constant, a shift in the plasmon resonance leads to a change in the intensity of the reflected beam. The shift is restricted locally and happens only in areas where the optical properties have changed. The usage of a two-dimensional array detector, compared with a linear array detector, therefore allows measurement of an SPR image.

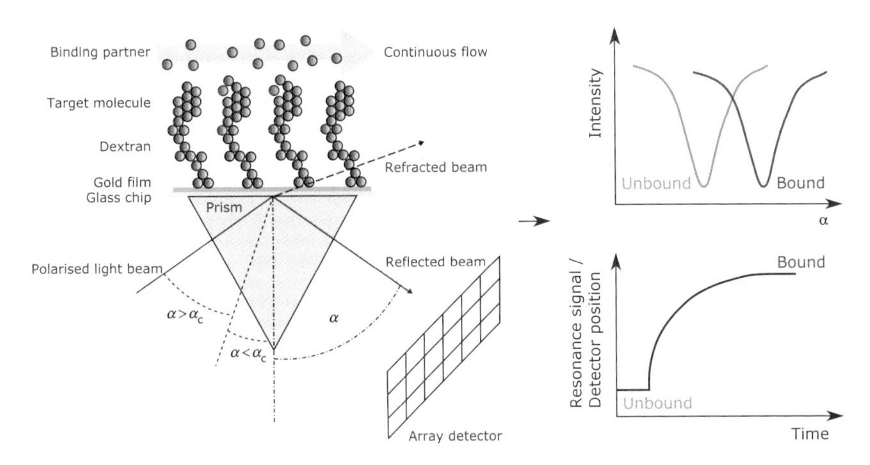

Fig. 5.1. Principle of SPR measurement. The angle α of total internal reflection shifts if material is bound on the sample side of the SPR chip. A time-dependent trace or sensorgram is recorded to monitor binding of material to the chip in the course of an experiment. Note that kinetic parameters can be determined from the sensorgram.

Surface plasmon resonance can detect changes in the refractive index of less than 10^{-4} or changes in layer heights of about 1 nm. This enables not only the detection of binding events between biomolecules but also binding at protein domains or changes in molecular monolayers with a lateral resolution of a few micrometres.

For SPR, light of wavelengths between infrared (IR) and near-IR (NIR) may be used. In general, the longer the wavelength of the light used, the better the sensitivity but the less the lateral resolution. In contrast, if high lateral resolution is required, red light needs to be used, as the propagation length of the plasmon wave is approximately proportional to the wavelength of the exciting light.

5.1.2 Applications

SPR (Biacore systems)

Generally, all two-component binding reactions can be investigated, which opens a variety of applications in the areas of drug design (protein–ligand interactions), membrane-associated proteins (protein–membrane binding), DNA-binding proteins, etc.

When investigating binding in a two-component system, one of the components involved needs to be tethered to the SPR chip. The tethering event itself needs to be validated to ensure that the component has been successfully immobilised and is present in a functional state.

Subsequently, the second component is added to the immobilised molecules by a continuous flow. If binding events take place, these can be followed in the association phase of the sensorgram and enable the determination of the on-rate constant (k_{on}). Typically, this is then followed by a washing step using buffer without the second component. This allows dissociation of the bound molecules over time and thus determination of the off-rate constant (k_{off}). From these data, the binding constants (association constant K_a and dissociation constant K_d) for the interaction can be calculated:

$$K_a = \frac{k_{on}}{k_{off}} = \frac{1}{K_d} \tag{5.1}$$

For biomolecular interactions, SPR can be used to determine very-low-affinity interactions. In drug discovery applications, SPR is commonly used as secondary screening to validate a compound 'hit' and determine how fast and stable the binding event is.

SPR imaging

The focus of SPR imaging experiments shifted in recent years from characterisation of ultrathin films to analysis of biosensor chips, especially affinity sensor arrays (Homola, 2003). The technique of SPR imaging can detect DNA–DNA, DNA–protein and protein–protein interactions in a two-dimensional manner. The detection limit for such biosensor chips is at the order of 10^{-9}–10^{-15} M (nM to fM). Apart from the detection of binding events as such, the quality of binding (low affinity, high affinity) can also be assessed by SPR imaging. Promising future applications for SPR imaging include peptide arrays that can be prepared on modified gold surfaces. This can prove useful for assessing peptide–antibody interactions. The current time resolution of less than 1 s for an entire image also allows high-throughput screening and *in situ* measurements.

5.2 Quartz crystal microbalance

The quartz crystal microbalance (QCM) allows surface-sensitive characterisation of processes by measuring the mass of a thin film adhering on a surface. Like SPR (see Section 5.1), QCM is a method for *in situ* studies of molecular adsorption and/or interaction processes on surfaces. This relatively simple technique allows the analysis of label-free biomolecules in direct, real time, and highly sensitive mass measurements in solution. In contrast to optical techniques such as SPR, QCM experiments measure the mass of an adsorbed film, including the solvent inside the film. The conventional QCM methodology has been advanced by the addition of dissipation monitoring (QCM-D) based on measurement of mass and viscoelastic properties of adsorbed films (Q-sense patent, 1995) (Rodahl *et al.*, 1996), leading to the current state-of-the-art QCM-D. In the life sciences, the main applications of QCM concern implants, biosensors, biochips and marine technologies.

5.2.1 Principle

Quartz crystal microbalance measurements rely on the piezoelectric properties of quartz crystals, which are the crystalline form of silica (SiO_2). The piezoelectric effect was discovered in 1880 by Pierre and Jacques Curie. If a piezoelectric crystal is subjected to mechanical pressure, an electric potential is induced across the crystal. Conversely, an electric potential across the crystal results in a mechanical deformation. Piezoelectric crystals are widely used in numerous applications, such as ultrasound, gas lighters, watches, atomic force microscopy (AFM) scanners (see Section 5.4) and QCM sensors.

A QCM sensor is a thin disc of quartz cut at a specific angle from a larger quartz crystal. Two metal electrodes situated on opposing sides of the crystal allow application of an electric potential across the crystal. An alternating voltage applied to the quartz crystal induces its deformation and results in a shear oscillation. Stimulated by the frequency of the applied voltage, the crystal can resonate at its fundamental resonance frequency or an overtone thereof. The measured resonance frequency is dependent on the total oscillating mass, which comprises the mass of the QCM sensor and that of the adsorbed film. Changes in mass on the QCM sensor are related to changes in frequency of the oscillating crystal through the Sauerbrey equation (Sauerbrey, 1959):

$$\Delta f_n = -\left(\frac{n}{C}\right)\Delta m_f = -\left(\frac{n}{C}\right)\Delta(\rho_f h_f) \qquad (5.2)$$

where Δf_n is the change of resonance frequency, n is the overtone order, Δm_f is the change of mass per unit area of the adsorbed film, and ρ_f and h_f are the density and the thickness of the adsorbed film, respectively. The mass sensitivity constant C depends on the fundamental resonance frequency and the material properties of the quartz crystal. For a 5 MHz quartz crystal, $C = 17.7$ ng Hz^{-1} cm^{-2}. The Sauerbrey equation is valid for adsorbed films that possess a small mass compared with the mass of crystal; the films must be evenly distributed and rigidly attached to the surface.

Upon binding of biomolecules to the film on the QCM sensor, a decrease in the resonance frequency is observed that is proportional to the mass of the resulting adsorbed film. The mass sensitivity is in the order of ng cm^{-2}. If the effective density ρ_f of the adsorbed film can be estimated, the film thickness h_f can be calculated.

In addition to the resonance frequency measurement, the dissipation factor D can be determined with QCM-D instruments. The D factor is defined as the ratio between the energy dissipated per cycle of oscillation and the total energy stored in the system, and provides information about the viscoelastic properties of the adsorbed film.

5.2.2 Applications

As QCM experiments can investigate the kinetics of mass changes (and simultaneously, in the case of QCM-D, structural changes) of the adhering film on the QCM sensor, this technique has become widely used in laboratories for the study of interfaces and interface processes in bulk solutions. Numerous applications have been reported, including analysis of biomaterials (for example, lipid vesicle interactions on surfaces) or membrane-associated proteins (protein–membrane binding), drug design (for instance, by analysing protein–ligand interactions) and control of protein binding or release.

For example, the QCM-D technique has been used to monitor the real-time kinetics of processes undergone by lipid vesicles on solid substrates (Keller & Kasemo, 1998). Lipid vesicles adsorb to solid surfaces and can form either a supported vesicular layer or a supported lipid mono- or bilayer. The QCM-D studies revealed the mechanisms of the vesicle fusion process leading to the formation of supported lipid mono- or bilayers, and characterised the dependence of these processes on factors such as temperature, vesicle concentration and vesicle lipid composition.

Existing diagnostic techniques can be transformed into a QCM application. For instance, the protein annexin A5 with its established functionalisation abilities can be coupled to a QCM sensor and has been used as a prototype protein to develop a platform for cell adhesion detection (Berat *et al.*, 2007). In a recent application, an annexin A5-functionalised QCM sensor was used for the quantitative detection of early apoptosis (Pan *et al.*, 2013). Annexin A5 conjugated to a fluorescent dye is a widely used tool in apoptosis detection assays. The protein preferentially binds to the acidic phospholipid phosphatidyl serine, which is generally believed to be exposed on the extracellular surface of cells undergoing apoptosis.

The value of QCM-D for biological systems is illustrated by a study investigating models for cell membranes; more realistic models consider the presence of conjugated lipids rather than pure lipid membranes. In this context, the formation of a hyaluronan film immobilised on a supported lipid bilayer was investigated as a mimic for pericellular coats of cell membranes (Richter *et al.*, 2007). Hyaluronan is a ubiquitous linear disaccharide in the extracellular matrix. In this study, a supported lipid bilayer containing a fraction of biotinylated lipids was formed on a solid surface. A solution of streptavidin was passed over the supported lipid bilayer and the binding was measured by QCM-D, indicating the formation of a stable monolayer of streptavidin. The various end-biotinylated hyaluronan solutions were then applied and hyaluronan was immobilised onto the supported lipid bilayer via streptavidin. Assisted by QCM-D analysis, the quantity of the immobilised hyaluronan can be well controlled, and the viscoelastic properties of such films, as well as their susceptibility to biotinylated probes, can be assessed.

5.3 Monolayer adsorption

5.3.1 The Langmuir–Blodgett method

A Langmuir film is made of insoluble molecules with a well-controlled packing density at the interface between air and a liquid subphase composed mostly of water. The theory of such films, as well as the experimental approach, was established by Langmuir in 1915. A few years later, Langmuir and Blodgett demonstrated the possibility of depositing multi-layer films onto solid substrates, a technique that became known as the Langmuir–Blodgett method.

Important applications of these films include investigation of physical and chemical properties of biomolecules and biosensors. The films are accepted models of biological membranes in the investigation of membrane interactions, and are thus of eminent importance for studying the interactions of drugs and active molecules with membranes.

Principle

In order for a molecule to be a surface-active agent (surfactant), it must possess amphiphilic properties, meaning that it consists of a hydrophilic (water-soluble) and a hydrophobic (water-insoluble) moiety. With the help of a volatile and water-insoluble solvent (e. g. chloroform), these surfactants can easily be spread over an aqueous buffer (called a subphase) contained in the Langmuir trough and self-arrange to form an insoluble

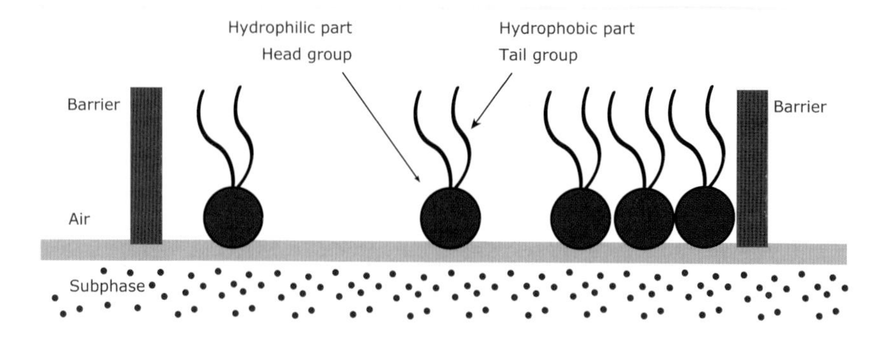

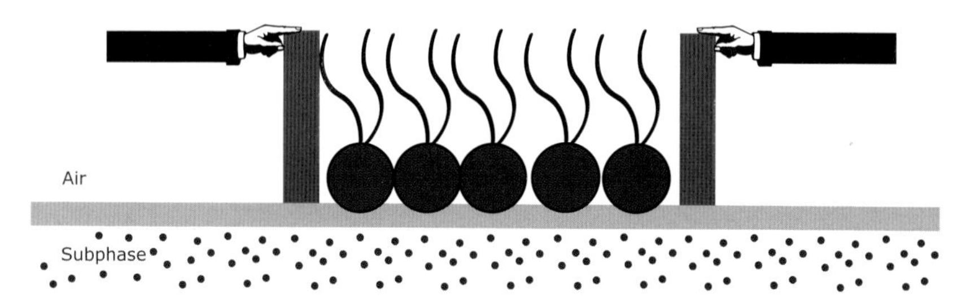

Fig. 5.2. Amphiphilic molecules spreading at an air–water interface. Depending on the available area, weak or strong interactions between individual molecules are observed.

monolayer – a Langmuir film. In this film, the polar head groups of the amphiphilic molecules are oriented towards aqueous subphase and the hydrophobic tails are turned towards the air (Fig. 5.2).

Prominent examples of important surfactants are phospholipids, which are found in biological membranes. The hydrophobic moiety of phospholipids comprises fatty acid chains of no less than 12 hydrocarbon groups. The carboxylic acids are conjugated to glycerol via ester bonds, thereby forming the diacylglycerol backbone. The remaining alcohol group is esterified with the phosphate ester of a polar 'head group' such as serine, ethanolamine or choline (Fig. 5.3), which thus constitutes the hydrophilic moiety. Phospholipids possessing two carbon chains of equal length are called diacyl phospholipids.

At an interface between two phases (e.g. at an air–water interface), the surface tension is a cohesive phenomenon arising from the attraction of solvent molecules towards each other on the liquid's surface. Conceptually, the surface tension can be regarded as force per distance on the surface. Polar liquids, such as water, have strong intermolecular interactions and thus high surface tensions. Any factor that decreases the strength of these interactions, for instance an increase in temperature

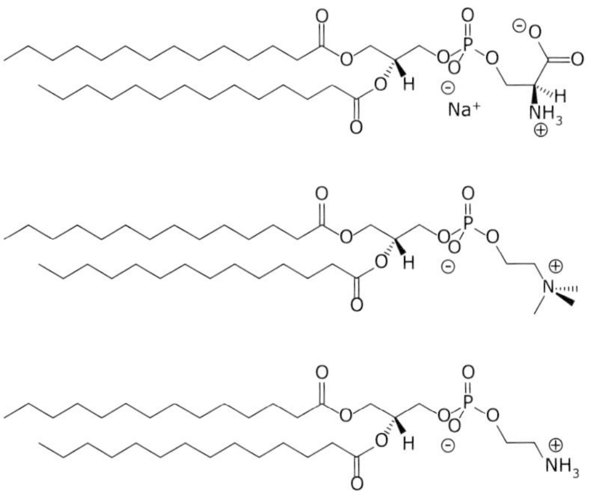

Di-myristoyl-phosphatidylserine (DMPS)

Di-myristoyl-phosphatidylcholine (DMPC)

Di-myristoyl-phosphatidylethanolamine (DMPE)

Fig. 5.3. Structures of commonly used phospholipids in Langmuir–Blodgett monolayer films.

or the addition of surfactants, will lower the surface tension. The change in the surface tension is called lateral surface pressure and is given by:

$$\pi = \gamma_0 - \gamma. \tag{5.3}$$

Here, π is the surface pressure, γ_0 is the surface tension of the pure subphase and γ is the surface tension of the subphase with the adsorbed film. The lateral surface pressure is the two-dimensional analogue of the pressure and has the unit $[\pi] = 1\ \mathrm{N\ m^{-1}}$. The surface pressure is generally measured with a Wilhelmy plate, a thin plate made of platinum or chromatography paper, partially immersed into the subphase.

When the area available for the surfactant molecules in the Langmuir trough is large, the distance between adjacent molecules is large and their interactions are weak. The surface tension of the subphase is thus only marginally reduced, and the measured surface pressure is small. If the available surface area of the monolayer is reduced by a barrier system, the film is compressed and the surfactant molecules start to exert a repulsive effect on each other. This significantly decreases the surface tension of the subphase and results in a high surface pressure (Fig. 5.2).

Plotting the surface pressure π as a function of the surface area A of the monolayer in the Langmuir trough produces a graph known as a pressure–area isotherm (Fig. 5.4). Analysis of these isotherms provides information on the arrangement of the surfactants at the interface. In the case of phospholipids, at constant temperature, this curve often contains sharp bends or kinks indicative of phase transitions in the two-dimensional layer.

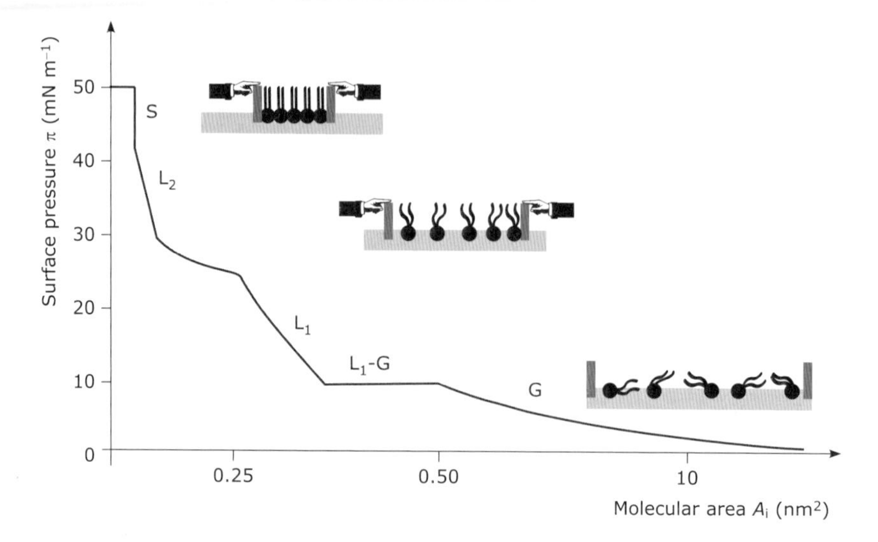

Fig. 5.4. Phase transition upon compression of a lipid monolayer. At low pressure (large area), the molecules are present in the gas phase (G). Upon a decrease of the area, the monolayer is compressed, the surface pressure increases and the molecules are forced into a liquid (L) and subsequently solid state (S). A further decrease of the area causes the collapse of the monolayer.

The phase behaviour of the monolayer is determined mainly by the physical and chemical properties of the surfactants, the subphase temperature and the subphase composition. A simple terminology used to classify different monolayer phases of fatty acids was proposed by Harkins (1952). When a large area is available, the monolayers exist in the gaseous state (G) and, upon compression, can undergo a phase transition to the liquid-expanded phase (L_1). Upon further compression, the L_1 phase undergoes a transition to the liquid-condensed phase (L_2), and at even higher densities the monolayer finally reaches the solid phase (S). If the monolayer is further compressed after reaching the solid phase, it will collapse into three-dimensional structures. The collapse is generally seen as a rapid decrease in the surface pressure or as a horizontal break in the isotherm (Fig. 5.4). Critical points in a surface pressure–area isotherm are:

- the molecular area A_i at which an initial pronounced increase in the surface pressure is observed;
- the surface pressure at which the phase transition occurs between the L_1 and L_2 phase; and
- the surface pressure at which the phase transition occurs between the L_2 and S phase.

Applications

Because of the intrinsic complexity of physiological membranes, the choice of a simplified model membrane, where all experimental parameters can be accurately monitored, is of great importance in the study of the fundamental processes of protein–lipid interactions. Phospholipid monolayers, prepared by the Langmuir–Blodgett technique, are the best-established model systems of biological membranes. The investigation of the phase behaviour and molecular organisation of lipid monolayers is a valuable tool for the study of interfacial properties of membrane-associating proteins and peptides.

For applications in structural biology and analysis of protein properties, two experimental settings are used:

- Recording of surface pressure–area isotherms. Isotherms are used to study changes in the phase diagrams that occur upon insertion of proteins into the lipid monolayer.
- Recording the change of pressure while keeping a constant area (area control). These experiments are designed to monitor changes in surface pressure when a protein inserts into the monolayer.

The Langmuir–Blodgett technique is of further particular importance for the preparation of thin lipid films, as it allows the deposition of monolayers on almost any kind of solid substrate. The technique provides precise control of the monolayer thickness, allows homogeneous deposition of the monolayer over large areas and also enables generation of multi-layer structures with varying layer composition.

5.3.2 Brewster angle microscopy

A Brewster angle microscope is based on the phenomenon of light reflectivity at the interface of two media: when an air–water interface is illuminated with a p-polarised light beam (Fig. 5.5) at a specific angle of incidence known as the Brewster angle (θ_B), then there is no reflection. The presence of surfactant at the air–water interface modifies the Brewster angle and the light beam is reflected if the angle of incidence has been set to the Brewster angle of the air–water interface without surfactant.

The technique was introduced independently in 1991 by Hönig and Möbius (Hönig & Möbius, 1991) and by Hénon and Meunier (Hénon & Meunier, 1991). Current state-of-the-art Brewster angle microscopes possess a spatial resolution of around 1–2 μm, which makes Brewster angle microscopy a unique tool to visualise surfactant distribution, their

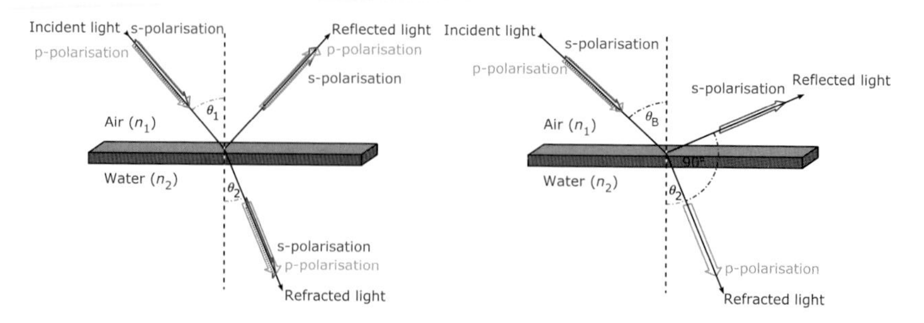

Fig. 5.5. Left: Incident light hits an air–water interface at any angle θ_1 (measured between the direction of the incident light and the normal of the interface surface), which results in reflected and refracted light beams. All polarisation components are conserved. Right: If the incident light hits the surface at the Brewster angle θ_B, the reflected light will be s-polarised and the refracted light will be p-polarised.

structural organisation (often in form of clusters called domains), their size and their distribution at the air–water interface. The technique enables observation of samples in real time and does not require labelling of surfactants.

Principle

A light beam impinges on an interface between one medium and another (e.g. air and water) with an angle θ_1 made by the incident light with the normal to the surface. Light is reflected with an angle $-\theta_1$ (made by the reflected light with the normal at surface) and also refracted with an angle θ_2 (made by the refracted light with the normal to the surface) defined by the Snell–Descartes law as:

$$n_1 \sin\theta_1 = n_2 \sin\theta_2 \tag{5.4}$$

where n_1 and n_2 are the refractive indices of the two media.

We have already seen in Section 1.1 that light can be described as electromagnetic radiation with the $\vec{E}$ and $\vec{M}$ vectors perpendicular to the direction of propagation (see Fig. 1.2). The direction of the electric field vector defines the polarisation angle of the light. In Fig. 5.5, p-polarisation means that the electric field vector $\vec{E}$ is oriented parallel to the plane of incidence, which is defined by the incident beam and the normal of the surface. Conversely, s-polarisation is achieved if the electric field is oriented perpendicular to this plane. If the incident light has been either p- or s-polarised, then the reflected and refracted beams can be observed, with exactly the same polarisation, under almost all incident angles θ_1.

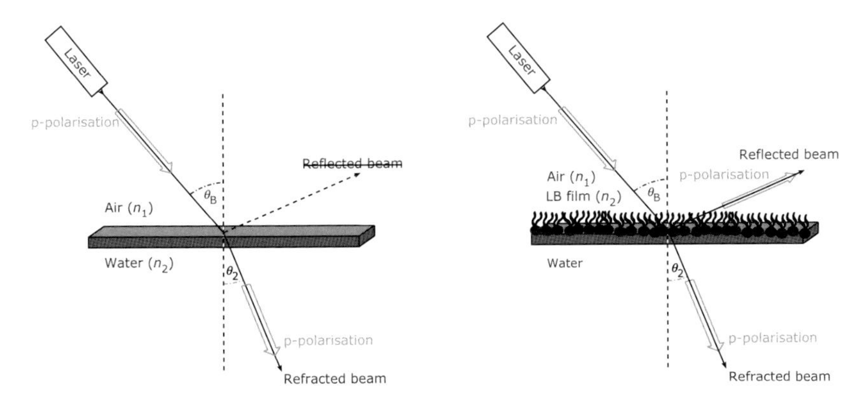

Fig. 5.6. Schematics of Brewster angle microscopy on a Langmuir–Blodgett film.

There is one notable exception, however. Brewster discovered in 1815 that there exists a particular angle of incidence where the reflected light is s-polarised and the refracted light is p-polarised. This critical angle is called the Brewster angle θ_B. This phenomenon forms the basis of Brewster angle microscopy (Fig. 5.6).

Brewster angle microscopy uses p-polarised light provided by a laser. With the light striking an air–water interface at an incident angle of θ_B, there is no reflected beam observed, and the refracted light beam is polarised. The Brewster angle can be derived from the Snell–Descartes equation and the Fresnel conditions (which are beyond the scope of this text), and depends only on the refractive indices of the two interfacing media:

$$\tan \theta_B = \frac{n_2}{n_1} \tag{5.5}$$

For the air–water interface, the Brewster angle is $\theta_B \approx 53°$, as the refractive indices of air and water are $n_1 = 1.00$ and $n_2 = 1.33$, respectively. At the Brewster angle, the reflectivity is near zero, and a dark image without any contrast is observed. Spreading of surfactant at the air–water interface (Fig. 5.6) induces a change in the refractive index n_2 and thus of the Brewster angle. As the angle of incidence of the light striking the interface is kept constant, a reflected beam is detected and bright images can be observed.

Applications

Brewster angle microscopy is a valuable tool for the study of a wide range of molecules at the air–water interface. In particular, combined with the Langmuir technique (see Section 5.3.1), it provides a wealth of details about the two-dimensional structure of lipid films. From Brewster angle microscopy images, structural information about the lipid orientation in, and formation of, ultrastructural domains can be obtained. When compared over time, Brewster angle microscopy images of lipid monolayers in the absence and presence of biomolecules such as peptides or proteins yield clues as to structural mechanisms of membrane association. It is also possible to estimate the film thickness from such images.

5.4 Atomic force microscopy

Atomic force microscopy (AFM) is a surface-sensitive characterisation method for imaging surfaces or detecting interactions between molecules. The surface is visualised in three dimensions by touching it with a tiny probe called the tip.

The method was conceived by Binnig and colleagues in the mid-1980s as a further development of scanning tunnelling microscopy, which they had established earlier. Atomic force and scanning tunnelling microscopy belong to the family of scanning probe microscopy (SPM). Unlike traditional microscopic techniques, SPM does not use glass or electromagnetic lenses to focus light or an electron beam. Rather, SPM technologies use the physical elevation of surfaces (z direction) as the parameter to be probed. In the case of AFM, the physical phenomenon probing surface elevation is the interaction force between the tip and the sample surface.

The most important advantage of AFM compared with other SPM microscopies is its applicability to a wide variety of samples. It can be used to probe conductor or non-conductor surfaces, and it can operate in vacuum, air or liquid. This allows testing of samples under physiologically relevant conditions, in buffer solution, at ambient temperature and without requiring fixation, staining or labelling. The technique of AFM was first applied to proteins in their aqueous environment by Hansma and colleagues in 1989 (Drake *et al.*, 1989), and has become very popular in many fields including materials science, polymer science, physics, life sciences and nano-biotechnology.

5.4.1 Principle

Atomic force microscopy provides an image of a sample's surface topography at high resolution. The surface area (x and y directions) is scanned in a raster-like fashion with a tip brought very close to the sample. The tip is attached to the free end of a cantilever (Fig. 5.7), which will be bent due to interactions between the tip and the sample surface.

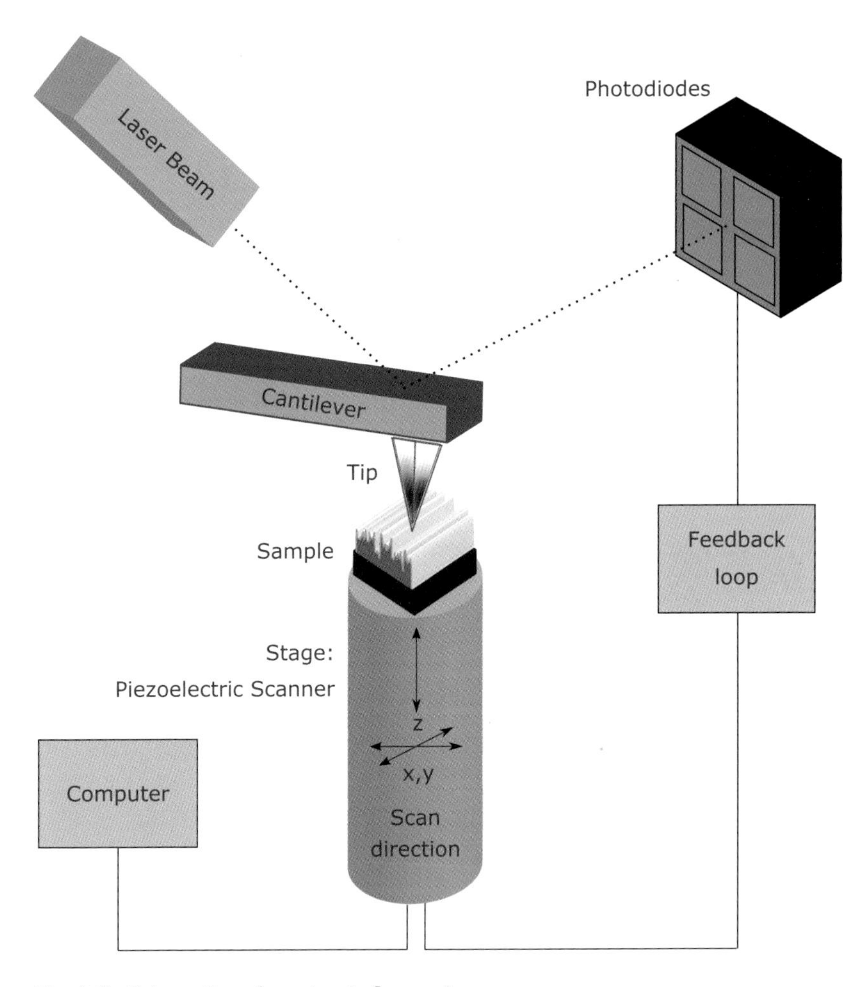

Fig. 5.7. Schematics of an atomic force microscope.

Attractive or repulsive forces between tip and surface (see Section 5.4.3) are observed through measurement of the reflection of a laser beam by the cantilever. The reflected laser beam is detected by a photodiode detector, which converts the optical signal into an electrical signal. The microscope uses a feedback loop, which triggers a z-movement of the piezoelectric scanner upon bending of the cantilever. The parameters of the feedback loop are optimised so as to minimise and maintain a constant value of the force between the tip and the sample. Scanning of the sample is achieved by x- or y-translation of the piezoelectric stage. The data are then assembled to obtain a map of sample surface height for a particular area.

The lateral resolution is dominated by the diameter and geometry of the tip apex. With current instruments, high-resolution images at a lateral resolution of 0.5–1 nm can be obtained. The vertical resolution is higher but is limited by mechanical vibrations and thermal fluctuations of the cantilever. Typically, vertical resolutions of about 0.1–0.2 nm can be achieved.

5.4.2 Modes of operation

An atomic force microscope can be operated in different imaging modes, depending on the type of sample to be investigated.

Contact mode

In this mode, the tip is in direct physical contact with the sample, and the instrument measures the repulsive forces between the tip and the sample during the scan. The repulsive force, and thus the deflection of the cantilever, is kept constant by the feedback loop, thus giving rise to vertical movements of the piezoelectric stage. This motion of the piezoelectric stage in the z direction is recorded as a height image.

With biological samples, this mode can lead to disruption of the tip and/or the sample. However, this mode can be used with many different types of biological samples. Additional factors such as the absence of sample preparation and a wide range of observation areas (ranging from μm^2 to nm^2) make this a very popular operation mode.

Intermittent contact mode (or Tapping® mode)

Here, the cantilever is oscillating close to its resonance frequency, and the tip periodically touches the surface. The amplitude and phase of the oscillation change due to forces acting between the tip and the sample. In this mode, the feedback loop is used to keep the amplitude constant, which gives rise to vertical displacements of the piezoelectric stage. The motion of the stage in the z direction is recorded as a height image, and phase changes of the tip oscillation are recorded in a phase image.

This mode is used with soft and fragile samples.

Non-contact mode

In the non-contact mode, the tip always stays at a distance above the sample surface. As with the intermittent contact mode, the cantilever is oscillating close to its resonance frequency, and the oscillation amplitude and phase are recorded as described above. In contrast to the intermittent contact mode, the oscillation phase of the tip changes and is readjusted through the feedback loop.

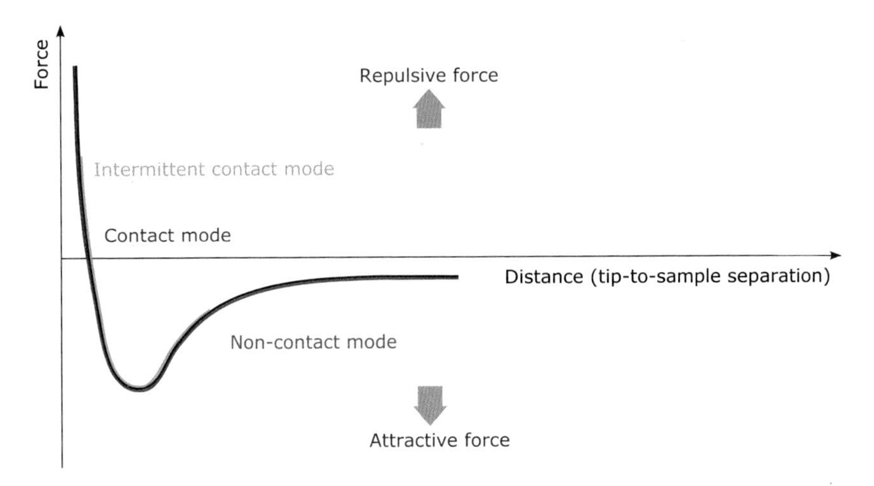

Fig. 5.8. Overview of forces exerted between the tip and the sample in AFM.

As the interactions of tip and sample are minimised in the non-contact mode, even very soft samples do not suffer any damage from the tip.

5.4.3 Forces

Atomic force microscopy operates by measuring attractive or repulsive forces between a tip and a surface. The forces encountered by a tip during AFM engagement in air are illustrated in Figure 5.8 and are due to the following molecular interactions:

- Adhesion forces (attractive)
- Electrostatic interactions (attractive or repulsive)
- Van der Waals interactions (attractive)
- Born repulsion (repulsive)
- Elastic deformation (attractive)
- Capillary force, if AFM is done in air (attractive).

The AFM tip experiences an increasingly attractive force as it gets into the spatial vicinity of the sample surface. At very short distances, repulsive forces counteract the attraction and dominate once the tip is in contact with the surface. In contact mode, the tip is in direct physical contact with the sample; the force between the tip and the sample is a repulsive force. In intermittent contact mode and in non-contact mode, the tip oscillates and either touches the sample briefly (intermittent contact mode) or sways above the sample surface (non-contact mode).

5.4.4 Instrumentation

Probe

As described above, the AFM probe consists of a cantilever with a sharp tip at one end. The cantilever operates very much like a spring: it is characterised by the spring constant and resonant frequency. The deflection of the cantilever (x) is proportional to the force acting on the tip, according to Hooke's law:

$$F = -k \times x, \tag{5.6}$$

where k is the spring constant of the cantilever.

The probe is chosen according to the sample to be studied. For biological samples, for instance, a cantilever with a low spring constant and a high resonance frequency is essential in order to reduce damage of the sample and enable quick acquisition of data. Typically, the spring constant of the cantilever is less than 0.1 N m^{-1} and the resonant frequency is higher than 200 kHz. Tips are available in different shapes (rectangular or V-shape) and specifications (length, width and thickness). Generally, they are made of silicon (Si) or silicon nitride (Si_3N_4). A thin gold layer may be deposited on the upper side of the tip in order to enhance reflectivity of the laser beam. The geometry and diameter of the tip apex limit the lateral resolution of AFM.

Piezoelectric scanner

The sample is mounted on the piezoelectric stage (scanner), which laterally displaces either the sample or the tip, depending on the instrument, in the x- and y-directions and controls the tip–sample interactions by vertical adjustments of the sample in the z direction. Due to the piezoelectric properties of the stage (see Section 5.2.1), an applied voltage leads to an expansion or a contraction, which is translated into an x-, y- or z-displacement of the stage. An AFM image corresponds to a series of scan lines in the y-direction. Each pixel along the line is a plot of voltage applied in the x-direction of the stage versus voltage applied to its z direction.

Laser detection system and feedback loop

When scanning over a surface, the cantilever bends, and the piezoelectric stage is moved vertically to adjust the forces between the probe and the surface to a constant value. To detect displacement of the cantilever, a laser beam is reflected from the back of the cantilever and detected by a

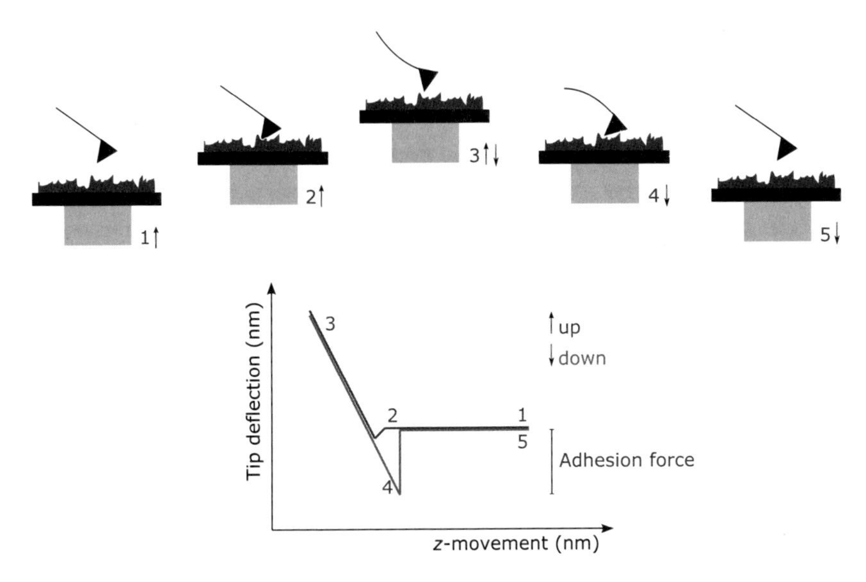

Fig. 5.9. Schematic illustration of a force–distance curve experiment at fixed x and y. The plot shows the correlation of the cantilever deflection and the movement of the piezoelectric stage in the z direction. The adhesion force is the attractive force between the tip and the sample surface.

photodiode array. The deflection of the cantilever is thus converted into an electrical signal.

The feedback loop integrates the detector signal and applies a feedback voltage to the z direction of the piezoelectric scanner to balance the cantilever deflection and thus regulate the force between the tip and the sample.

5.4.5 Force–distance curves

The basis of force spectroscopy is constituted by the so-called force–distance curve, which can be determined by AFM. It is a plot of the interaction forces between tip and sample (measured as cantilever deflection) versus tip–sample distance. In order to obtain such curves, the piezoelectric stage is ramped along the z direction in a cyclic up and down movement (Fig. 5.9). The tip–sample force can be calculated by Hooke's law from the recorded cantilever deflection.

Force–distance curves show five distinct phases as depicted in Fig. 5.9:

1. The piezoelectric stage with the sample approaching the tip.
2. The tip is brought close to the sample surface, but there is no contact between tip and surface. An attractive force between the tip and the

surface induces a jump of the cantilever, leading to contact between the tip and the surface.

3. The deflection signal increases as the piezoelectric stage and the sample move upwards with the tip being in contact with the surface.
4. The piezoelectric stage moves the sample down and the deflection signal decreases. Due to an attractive force between the tip and the sample surface, the deflection continues to decrease beyond the z-distance where contact was established in the upwards movement. The cantilever is bent down as the piezoelectric stage retracts.
5. The tip retracts from the surface as the stage continues on the downwards z-movement. The cantilever jumps back into its idle position. The force difference at the point of rupture is called the adhesion force and corresponds to the force required to overcome attractive forces between the tip and the surface.

5.4.6 Applications

AFM imaging

With AFM, all biomolecules can be imaged in air, liquid, native or reconstituted conditions. Thus, AFM studies have been conducted that image DNA (Moukhtar *et al.*, 2010), membrane proteins (Bahatyrova *et al.*, 2004), lipid bilayers (Grandbois *et al.*, 1998) and cells (Simon *et al.*, 2003). High-resolution images of biomolecules can be provided by AFM imaging, and it can in principle follow real-time dynamic process *in situ*. Time resolution with conventional AFM instruments has been limited, but first results with high-speed AFM (see Noi *et al.*, 2013; Rico *et al.*, 2013, in Section 5.5) indicate promising new applications of this technique.

Force spectroscopy

Force spectroscopy is used mainly to conduct force–distance curve experiments that yield information about forces between single molecules or elastic properties of a sample (e.g. cells). For example, an AFM tip was used to measure forces that are related to the removal of single peptide molecules from a bilayer (Ganchev *et al.*, 2004). A wide range of systems can be investigated by this technique when the AFM probe is individually functionalised.

When using functionalised probes, where for example antibodies are covalently linked to the tip via a long flexible spacer group (Hinterdorfer *et al.*, 1996), force spectroscopy allows characterisation of specific interactions between antibody and antigen, as well as measurements of forces between chemical groups, receptor protein and ligand, antibody and antigen, and molecular assemblies, depending on the functionalisation of the probe.

FURTHER READING

Surface plasmon resonance

Anker, J. N., Hall, W. P., Lyandres, O. *et al.* (2008). Biosensing with plasmonic nanosensors. *Nature Materials* **7**, 442–53.

Campbell, C. T. & Kim, G. (2007). SPR microscopy and its applications to high-throughput analyses of biomolecular binding events and their kinetics. *Biomaterials* **28**, 2380–92.

Majka, J. & Speck, C. (2007). Analysis of protein–DNA interactions using surface plasmon resonance. *Advances in Biochemical Engineering/ Biotechnology* **104**, 13–36.

Neumann, T., Junker, H. D., Schmidt, K. & Sekul, R. (2007). SPR-based fragment screening: advantages and applications. *Current Topics in Medicinal Chemistry* **7**, 1630–42.

Phillips, K. S. & Cheng, Q. (2007). Recent advances in surface plasmon resonance based techniques for bioanalysis. *Analytical and Bioanalytical Chemistry* **387**, 1831–40.

Websites on surface plasmon resonance

BioNavis: Surface Plasmon Resonance basics: http://www.bionavis.com/technology/spr/

UK Surface Analysis Forum: An Introduction to SPR: http://www.uksaf.org/tech/spr.html

Biacore Life Sciences: Label-free interaction analysis: http://www.biacore.com/lifesciences/index.html

Quartz crystal microbalance

Cho, N.-J., Frank, C. W., Kasemo, B. & Höök, F. (2010). Quartz crystal microbalance with dissipation monitoring of supported lipid bilayers on various substrates. *Nature Protocols* **5**, 1096–106.

Website on quartz crystal microbalance

QCM-D animation: http://www.q-sense.com/media/player.aspx?file=/$2/file/q-sense-qcm-d-animation-2009-with-measurement-examples.flv

Monolayer adsorption

Calvez, P., Demers, E., Boisselier, E. & Salesse, C. (2011). Analysis of the contribution of saturated and polyunsaturated phospholipid monolayers to the binding of proteins. *Langmuir* **27**, 1373–9.

Maget-Dane, R. (1999). The monolayer technique: a potent tool for studying the interfacial properties of antimicrobial and membrane-lytic peptides and their interactions with lipid membranes. *Biochimica et Biophysica Acta – Biomembranes* **1462**, 109–40.

Volinsky, R., Kolusheva, S., Berman, A. &, Jelinek, R. (2006). Investigations of antimicrobial peptides in planar film systems. *Biochimica et Biophysica Acta – Biomembranes* **1758**, 1393–407.

Websites on monolayer adsorption

Dr Irving Langmuir air–water interface demonstration: http://www.youtube.com/watch?v=5sGUqX7EcBA

Katherine Blodgett – Engineering Pioneer: http://www.youtube.com/watch?v=epnHUNngVz0

Polarized Light Microscopy Interactive Java Tutorials: Brewster's Angle: http://micro.magnet.fsu.edu/primer/java/polarizedlight/brewster/index.html

Atomic force microscopy

Binnig, G., Roher, H., Gerber, C. & Weibel, E. (1982). Surface studies by scanning tunneling microscopy. *Physical Review Letters* **49**, 57–61.

Binnig, G., Quate, C. F. & Gerber, C. (1986). Atomic force microscope. *Physical Review Letters* **56**, 930–3.

Muller, D. J. (2008). AFM: a nanotool in membrane biology. *Biochemistry* **47**, 7986–98.

Muller, D. J. & Engel, A. (1997). The height of biomolecules measured with the atomic force microscope depends on electrostatic interactions. *Biophysical Journal* **73**, 1633–44.

Noi, K., Yamamoto, D., Nishikori, S. *et al.* (2013). High-speed atomic force microscopic observation of ATP-dependent rotation of the AAA+ chaperone p97. *Structure* **21**, 1992–2002.

Rico, F., Gonzalez, L., Casuso, I., Puig-Vidal, M. & Scheuring, S. (2013). High-speed force spectroscopy unfolds titin at the velocity of molecular dynamics simulations. *Science* **342**, 741–3.

Websites on atomic force microscopy

High-speed AFM: http://www.se.kanazawa-u.ac.jp/bioafm_center/HS-AFM_Div.htm

Development of a state-of-the-art single-molecule force-clamp spectrometer: http://fernandezlab.biology.columbia.edu/research/development-state-art-single-molecule-force-clamp-spectrometer

REFERENCES

Bahatyrova, S., Frese, R., Siebert, C. *et al.* (2004). The native architecture of a photosynthetic membrane. *Nature* **430**, 1058–62.

Berat, R., Remy-Zolghadry, M., Gounou, C. *et al.* (2007). Peptide-presenting two-dimensional protein matrix on supported lipid bilayers: an efficient platform for cell adhesion. *Biointerphases* **2**, 165–72.

Bettio, A. & Beck-Sickinger, A. (2001). Biophysical methods to study ligand-receptor interactions of neuropeptide Y. *Biopolymers* **60**, 420–37.

Bohr, N. (1913a). On the constitution of atoms and molecules. *Philosophical Magazine* **26**, 857.

(1913b). On the constitution of atoms and molecules. *Philosophical Magazine* **26**, 476.

(1913c). On the constitution of atoms and molecules. *Philosophical Magazine* **26**, 1–25.

Boute, N., Jockers, R. & Issad, T. (2002). The use of resonance energy transfer in high-throughput screening: BRET versus FRET. *Trends in Pharmacological Sciences* **23**, 351–4.

Boutet, S., Lomb, L., Williams, G. J. *et al.* (2012). High-resolution protein structure determination by serial femtosecond crystallography. *Science* **337**, 362–4.

Bragg, W. H. & Bragg, W. L. (1913). The reflexion of X-rays by crystals. *Proceedings of the Royal Society A* **88**, 428–38.

Brakmann, S. & Nöbel, N. (2003). FRET in der Biochemie. *Nachrichten aus Chemie, Technik und Laboratorium* **51**, 319–23.

Bundle, D. & Siguskjold, B. (1994). Determination of accurate thermodynamics of binding by titration microcalorimetry. *Methods in Enzymology* **247**, 288–305.

Chapman, H. N., Fromme, P., Barty, A. *et al.* (2011). Femtosecond X-ray protein nanocrystallography. *Nature* **470**, 73–7.

Cooper, A. (1999). Thermodynamic analysis of biomolecular interactions. *Current Opinion in Chemical Biology* **3**, 557–63.

Cooper, A. & McAuley, K. E. (1993). Microcalorimetry and the molecular recognition of peptides and proteins. *Philosophical Transactions of the Royal Society A* **345**, 23–35.

Drake, B., Prater, C. B., Weisenhorn, A. L. *et al.* (1989). Imaging crystals, polymers, and processes in water with the atomic force microscope. *Science* **243**, 1586–9.

Dunitz, J. (1995). Win some, lose some: Enthalpy-entropy compensation in weak intermolecular interactions. *Chemistry & Biology* **2**, 709–712.

Förster, T. (1948). Zwischenmolekulare Energiewanderung und Fluoreszenz. *Annals of Physics* **2**, 57–75.

Ganchev, D., Rijkers, D., Snel, M., Killian, A. & de Kruijff, B. (2004). Strength of integration of transmembrane alpha-helical peptides in lipid bilayers as determined by atomic force spectroscopy. *Biochemistry* **43**, 14 987–93.

Gill, S. C. & von Hippel, P. H. (1989). Calculation of protein extinction coefficients from amino acid sequence data. *Analytical Biochemistry* **182**, 319–26.

Giordano, L., Jovin, T., Irie, M. & Jares-Erijman, E. (2002). Diheteroarylethenes as thermally stable photoswitchable acceptors in photochromic fluorescence resonance energy transfer (pcFRET). *Journal of the American Chemical Society* **124**, 7481–9.

Grandbois, M., Clausen-Schaumann, H. & Gaub, H. (1998). Atomic force microscope imaging of phospholipid bilayer degradation by phospholipase A2. *Biophysical Journal* **74**, 2398–404.

Harkins, W. D. (1952). *The Physical Chemistry of Surface Films*. New York: Reinhold.

Hénon, S. & Meunier, J. (1991). Microscope at the Brewster angle: direct observation of first-order phase transitions in monolayers. *Review of Scientific Instruments* **62**, 936–9.

Heyduk, T. & Niedziela-Majka, A. (2002). Fluorescence resonance energy transfer analysis of *Escherichia coli* RNA polymerase and polymerase-DNA complexes. *Biopolymers* **61**, 201–13.

Hinterdorfer, P., Baumgartner, W., Gruber, H. J., Schilcher, K. & Schindler, H. (1996). Detection and localization of individual antibody–antigen recognition events by atomic force microscopy. *Proceedings of the National Academy of Sciences, USA* **93**, 3477–81.

Hofmann, A. & Wlodawer, A. (2002). PCSB – a program collection for structural biology and biophysical chemistry. *Bioinformatics* **18**, 209–10.

Holdgate, G. (2001). Making cool drugs hot: the use of isothermal titration calorimetry as a tool to study binding energetics. *BioTechniques* **31**, 164–84.

Homola, J. (2003). Present and future of surface plasmon resonance biosensors. *Analytical and Bioanalytical Chemistry* **377**, 528–39.

Hönig, D. & Möbius, D. (1991). Direct visualization of monolayers at the air-water interface by Brewster Angle Microscopy. *Journal of Physical Chemistry* **95**, 4590–2.

Kang, J., Piszczek, G. & Lakowicz, J. (2002). Enhanced emission induced by FRET from a long-lifetime, low quantum yield donor to a long-wavelength, high quantum yield acceptor. *Journal of Fluorescence* **12**, 97–103.

Keller, C. & Kasemo, B. (1998). Surface specific kinetics of lipid vesicle adsorption measured with a quartz crystal microbalance. *Biophysical Journal* **75**, 1397–402.

Kimura, C., Maeda, K., Hai, H. & Miki, M. (2002). Ca^{2+}- and s1-induced movement of troponin T on mutant thin filaments reconstituted with functionally deficient mutant tropomyosin. *Journal of Biochemistry* **132**, 345–52.

Klewpatinond, M. & Viles, J. H. (2007). Fragment length influences affinity for Cu^{2+} and Ni^{2+} binding to His96 or His111 of the prion protein and spectroscopic evidence for a multiple histidine binding only at low pH. *Biochemical Journal* **404**, 393–402.

Kohl, T., Heinze, K., Kuhlemann, R., Koltermann, A. & Schwille, P. (2002). A protease assay for two-photon crosscorrelation and FRET analysis based solely on fluorescent proteins. *Proceedings of the National Academy of Sciences, USA* **99**, 12 161–6.

Mach, H., Middaugh, C. R. & Lewis, R. V. (1992). Statistical determination of the average values of the extinction coefficients of tryptophan and tyrosine in native proteins. *Analytical Biochemistry* **200**, 74–80.

Morse, P. M. (1929). Diatomic molecules according to the wave mechanics. II. Vibrational levels. *Physics Review* **34**, 57–64.

Moshinsky, D. J., Ruslim, L., Blake, R. A. & Tang, F. (2003). A widely applicable, high-throughput TR-FRET assay for the measurement of kinase autophosphorylation: VEGFR-2 as a prototype. *Journal of Biomolecular Screening* **4**, 447–52.

Moukhtar, J., Faivre-Moskalenko, C., Milani, P. *et al.* (2010). Effect of genomic long-range correlations on DNA persistence length: from theory to single molecule experiments. *Journal of Physical Chemistry B* **114**, 5125–43.

Neutze, R., Wouts, R., van der Spoel, D., Weckert, E. & Hajdu, J. (2000). Potential for biomolecular imaging with femtosecond X-ray pulses. *Nature* **406**, 752–7.

Pan, Y., Shan, W., Fang, H. *et al.* (2013). Annexin-V modified QCM sensor for the label-free and sensitive detection of early stage apoptosis. *Analyst* **138**, 6287–90.

Perczel, A., Park, K. & Fasman, G. (1992). Analysis of the circular dichroism spectrum of proteins using the convex constraint algorithm: a practical guide. *Analytical Biochemistry* **203**, 83–93.

Popmintchev, T., Chen, M., Popmintchev, D. *et al.* (2012). Bright coherent ultrahigh harmonics in the keV x-ray regime from mid-infrared femtosecond lasers. *Science* **336**, 1287–91.

Remington, S. J. (2011). Green fluorescent protein: a perspective. *Protein Science* **20**, 1509–19.

Rhee, H., June, Y., Lee, J. *et al.* (2009). Femtosecond characterization of vibrational optical activity of chiral molecules. *Nature* **458**, 310–13.

Rice, P., Longden, I. & Bleasby, A. (2000). EMBOSS: The European Molecular Biology Open Software Suite. *Trends in Genetics* **16**, 276–7.

Richter, R. P., Hock, K. K., Burkhartsmeyer, J. *et al.* (2007). Membrane-grafted hyaluronan films: a well-defined model system of glycoconjugate cell coats. *Journal of the American Chemical Society* **129**, 5306–7.

Rodahl, M., Höök, F., Krozer, A., Brzezinski, P. & Kasemo, B. (1996). Quartz crystal microbalance setup for frequency and Q-factor measurements in gaseous and liquid environments. *Review of Scientific Instruments* **66**, 3924–30.

Rogers, M. S., Cryan, L. M., Habeshian, K. A. *et al.* (2012). A FRET-based high throughput screening assay to identify inhibitors of anthrax protective antigen binding to capillary morphogenesis gene 2 protein. *PLoS ONE* **7**, e3991.

Rosenblum, B., Lee, L., Spurgeon, S. *et al.* (1997). New dye-labeled terminators for improved DNA sequencing patterns. *Nucleic Acids Research* **25**, 4500–4.

Salzer, R. & Steiner, G. (2004). Oberflächenplasmonen-Resonanz in neuem Licht. *Nachrichten aus Chemie, Technik und Laboratorium* **52**, 809–11.

Sambrook, J., Fritsch, E. & Maniatis, T. (1989). *Molecular Cloning: a Laboratory Manual*. Cold Spring Harbor: Cold Spring Harbor Laboratory Press.

Sauerbrey, G. (1959). Verwendung von Schwingquarzen zur Wägung dünner Schichten und zur Mikrowägung. *Zeitschrift für Phyik* **155**, 206–22.

Schwartz, C. L., Heumann, J. M., Dawson, S. C. & Hoenger, A. (2012). A detailed, hierarchical study of *Giardia lamblia*'s ventral disc reveals novel microtubule-associated protein complexes. *PLoS ONE* **7**, e43783.

Simon, A., Cohen-Bouhacina, T., Porte, M. C. *et al.* (2003). Characterization of dynamic cellular adhesion of osteoblasts using atomic force microscopy. *Cytometry* **54A**, 36–47.

Song, L., Jares-Erijman, E. & Jovin, T. (2002). A photochromic acceptor as a reversible light-driven switch in fluorescence resonance energy transfer (FRET). *Journal of Photochemistry and Photobiology A: Chemistry* **150**, 177–85.

Stryer, L. & Haugland, R. (1967). Energy transfer: a spectroscopic ruler. *Proceedings of the National Academy of Sciences, USA* **58**, 719–26.

Trakselis, M., Alley, S., Abel-Santos, E. & Benkovic, S. (2001). Creating a dynamic picture of the sliding clamp during T4 DNA polymerase holoenzyme assembly by using fluorescence resonance energy transfer. *Proceedings of the National Academy of Sciences, USA* **98**, 8368–75.

Uhlemann, S., Müller, H., Hartel, P., Zach, J. & Haider, M. (2013). Thermal magnetic field noise limits resolution in transmission electron microscopy. *Physical Review Letters* **111**, 046101.

Ullman, E. F., Kirakossian, H., Singh, S. *et al.* (1994). Luminescent oxygen channeling immunoassay: Measurement of particle binding kinetics by chemiluminescence. *Proceedings of the National Academy of Sciences, USA* **91**, 5426–30.

Warburg, O. & Christian, W. (1941). Isolierung und Kristallisation des Gärungsferments Enolase. *Biochemische Zeitschrift* **310**, 384–421.

White, T. A., Kirian, R. A., Martin, A. V. *et al.* (2012). CrystFEL: a software suite for snapshot serial crystallography. *Journal of Applied Crystallography* **45**, 335–41.

Wiseman, T., Williston, S., Brandts, J. & Lin, L. (1989). Rapid measurement of binding constants and heats of binding using a new titration calorimeter. *Analytical Biochemistry* **179**, 131–7.

Xu, Y., Piston, D. & Johnson, C. (1999). A bioluminescence resonance energy transfer (BRET) system: application to interacting circadian clock proteins. *Proceedings of the National Academy of Sciences, USA* **96**, 151–6.

INDEX